Neurofilaments

Publications in the Health Sciences

Publication of this book was assisted by a
McKnight Foundation grant to the
University of Minnesota Press's program
in the health sciences and a MacArthur
Foundation Grant to C. A. Marotta.

Neurofilaments

Edited by
Charles A. Marotta
McLean Hospital,
Massachusetts General Hospital,
    and
Harvard Medical School

University of Minnesota Press, *Minneapolis*

Copyright ©1983 by the University of Minnesota.
All rights reserved.
Published by the University of Minnesota Press,
2037 University Avenue Southeast, Minneapolis, MN 55414
Printed in the United States of America.

Library of Congress Cataloging in Publication Data
Main entry under title:

Neurofilaments.

(Publications in the health sciences)
Bibliography: p.
Includes index.
1. Nervous system — Diseases — Addresses, essays,
lectures. 2. Cytoplasmic filaments — Addresses, essays,
lectures. I. Marotta, Charles A. II. Series.
[DNLM: 1. Cytoplasmic filaments. 2. Cytoplasmic
filaments — Pathology. 3. Neurofibrils. 4. Neurofibrils
— Pathology. WL 102.5 N4932]
RC347.N475 1983   616.8'047   83-6521
ISBN 0-8166-1254-4

Contents

Contributors

Fung-Chow Chiu
Department of Neurology, Albert Einstein College of Medicine, Bronx, New York 10461

Mark H. Ellisman
Department of Neurosciences, University of California, San Diego, School of Medicine, La Jolla, California 92093

James E. Goldman
Department of Pathology, Albert Einstein College of Medicine, Bronx, New York 10461

Khalid Iqbal
Department of Pathological Neurobiology, New York State Institute for Basic Research in Developmental Disabilities, 1050 Forest Hill Road, Staten Island, New York 10314

Kenneth S. Kosik
Ralph Lowell Laboratories, Mailman Research Center, McLean Hospital, Belmont, Massachusetts 02178, and Department of Neurology and Neuropathology, Harvard Medical School, Boston, Massachusetts 02115

Charles A. Marotta
Mailman Research Center, McLean Hospital, Belmont, Massachusetts 02178; Psychiatry Department, Massachusetts General Hospital, Boston, Massachusetts 02114; and Program in Neuroscience and Department of Psychiatry, Harvard Medical School, Boston, Massachusetts 02115

George S. Merz
Department of Pathological Neurobiology, New York State Institute for Basic Research in Developmental Disabilities, 1050 Forest Hill Road, Staten Island, New York 10314

Patricia A. Merz
Department of Pathological Neurobiology, New York State Institute for Basic Research in Developmental Disabilities, 1050 Forest Hill Road, Staten Island, New York 10314

Ralph A. Nixon
Ralph Lowell Laboratories, Mailman Research Center, McLean Hospital, Belmont, Massachusetts 02178, and Department of Psychiatry, Harvard Medical School, Boston, Massachusetts 02115

William T. Norton
Departments of Neurology and Neuroscience, Albert Einstein College of Medicine, Bronx, New York 10461

Alfred Pope
Ralph Lowell Laboratories, Mailman Research Center, McLean Hospital, Belmont, Massachusetts 02178, and Department of Neurology and Neuropathology, Harvard Medical School, Boston, Massachusetts 02115

Keith R. Porter
Department of Molecular, Cellular, and Developmental Biology, University of Colorado, Boulder, Colorado 80309

William W. Schlaepfer
Department of Pathology and Laboratory Medicine, University of Pennsylvania Medical School, Philadelphia, Pennsylvania 19104

Dennis J. Selkoe
Ralph Lowell Laboratories, Mailman Research Center, McLean Hospital, Belmont, Massachusetts 02178, and Department of Neurology and Neuropathology, Harvard Medical School, Boston, Massachusetts 02115

Guang Y. Wen
Department of Pathological Neurobiology, New York State Institute for Basic Research in Developmental Disabilities, 1050 Forest Hill Road, Staten Island, New York 10314

Mark Willard
Department of Anatomy and Neurobiology and Department of Biological Chemistry, Washington University School of Medicine, 660 South Euclid Avenue, St. Louis, Missouri 63110

Henryk M. Wisniewski
Department of Pathological Neurobiology, New York State Institute for Basic Research in Developmental Disabilities, 1050 Forest Hill Road, Staten Island, New York 10314

Editor's Note

The essays of *Neurofilaments* provide a detailed overview of an important area of investigation. The need for this volume was apparent since there existed no other book that presented in one source the most interesting and significant areas of research into neuronal intermediate filaments. Neurofilaments cannot be effectively considered without reference to neuronal macromolecular structure as well as function. Thus, chapters on the cytoskeleton and axonal transport were included to provide suitable background information. The contribution of normal neurofilaments to paired helical filaments is currently an unsettled area. This volume, however, appeared incomplete without a critical examination of neuronal filaments in the Alzheimer's disease brain. Neuroscience research historically has its roots in chemistry and physiology and more recently has borrowed extensively from cellular and molecular biology. These varied disciplines have rarely been combined more harmoniously and effectively than in current investigations on neurofilaments.

We wish to thank Leta Sinclair for serving as project coordinator; her persistent and selfless efforts are greatly appreciated. We are grateful to Dr. Richard M. Abel formerly of the University of Minnesota Press for his supportive and helpful comments during the preparation of this volume.

Historical Preface

Alfred Pope

Almost a century has passed since in 1886 Paul Ehrlich, using his techniques for intravital staining with methylene blue, demonstrated what had long been surmised—that nerve cell bodies contain intracytoplasmic fibrillae. The reduced silver impregnation methods of Bielschowsky (1902) and Ramon y Cajal (1903) showed that these "neurofibrils" course in all directions through the neuronal perikaryon, branching and anastomosing as they do and thus generating a true intracellular network. They extend as straight, parallel fibrils into both dendrites and axons to their end arborizations. That these objects are not an artifact of fixation was resolved by de Rénji's observations (1928) that showed their existence in living nerve cells and fibers.

The advent of neuroelectron microscopy in the mid-20th century (pioneered by Francis O. Schmitt and associates) revealed the presence, within both neuronal cytoplasm and its protoplasmic and axonal processes, of three types of fibrous cytoskeletal components. As in most cells, these include microtubules (approximately 240 Ångstrom [Å] units total diameter), intermediate filaments (100 Å), and microfilaments (50 Å) consisting of polymerized actin. Among these only the 100-Å intermediate filaments are argyrophilic. Therefore, the classical "neurofibrils" of light microscopy represent visualization of aggregates of intermediate (neuro-) filaments. These structures are the predominant cytoskeletal components of axons but are outnumbered in the perikaryon and dendrites by neurotubules. In spite of a wealth of information gained in recent years concerning the triplet of fibrous proteins that constitute the macromolecular substructure of neurofilaments, their biological role, other than as a structural support system, remains unknown, though involvement in axonal transport has repeatedly been suggested.

Pari passu with development of cytological knowledge concerning neurofibrils came recognition of their importance in neuropathology. In 1907, Alois Alzheimer reported neuropathological findings in the brain of a 51-year-old patient who had died following a progressive dementing illness. Alzheimer noted that, in addition to gross cerebral atrophy, Bielschowsky-reduced silver preparations revealed the presence of two characteristic histopathological lesions throughout the cerebral mantle. There were amorphous silver-staining intercellular deposits (the senile or "neuritic" plaques) and intraneuronal agyrophilic inclusions having multiple configurations but seemingly representing reduplication or fusion of the neurofibrils. These "neurofibrillary tangles," together with the argyrophilic plaques, do constitute the hallmark histopathological stigmata of the middle- and late-life dementia continuum comprising presenile Alzheimer's disease proper (Kraepelin's designation) and senile dementia of the Alzheimer type. However, these lesions are not confined to this class of dementing illnesses. They are a constant finding (especially in hippocampus) in the brains of elderly subjects who have not necessarily suffered significant cognitive impairment; evidently, therefore, they are an aspect of brain aging per se. In addition, they have been shown to occur in a dozen or more unrelated neurological diseases of diverse etiologies, which include genetic, infectious (viral), metabolic, and even traumatic antecedents. Indeed, tangle formation seems to be a relatively nonspecific neuronal response to a wide variety of neuropathological insults, and its pathogenesis, therefore, is a major challenge for neuroscientists interested in neurological disease.

Electron microscopy, especially in the hands of Robert Terry and associates, has again been of central importance in defining the cytological ultrastructure of the neurofibrillary tangle and senile plaque. Ultrastructurally, the tangle consists of orderly arrays of helically wound filament strands aptly described by the generic designation "paired helical filaments," as first proposed by Kidd. These are also found in the degenerating axonal telodendra that are a prominent feature of the neuritic (senile) plaque. The ordered arrangement of the paired helical filaments accounts for their characteristic birefringence, but, apart from their increased quantity and unusual configuration, they have thus far been found to be ultrastructurally indistinguishable from normal 100 Å filaments. Moreover, ascertainment of alterations in protein substructure that could account for these phenomena has remained a baffling problem, though intriguing leads are currently under active exploration.

During the past decade, interest in structural protein neurochemistry and its role in the study of brain aging and dementia has been intense. The central importance of the neurofilaments for these issues is self-evident, and an impressive number of creative investigations into their biology and pathology has been the result. For advancement of knowledge in this sphere, further illuminating research on the biochemical structure, biosynthesis, assembly, transport, and degradation of neurofilament proteins is an imperative. It is, therefore, commendable and timely that these focal concerns of neuroscientists worldwide be recognized and the state of the art authoritatively described as has been done in the articles comprising this volume. Such juxtapositioning of present knowledge and unanswered challenges concerning neurofilaments will surely help to generate new vistas for research and major advances in the understanding of the neurobiology and pathology of these decisive organelles.

Neurofilaments

Chapter 1

Introduction
to the Cytoskeleton

Mark H. Ellisman and Keith R. Porter

Introduction to the Organization of Cytoplasm

The concept of form-controlling components in the cytoplasm of eukaryotic cells predates the electron-microscopic description of fibrous macromolecules now thought to provide the cells' framework. Recognition that 24-nm microtubules (MT) (Figure 1-1) are a ubiquitous component of eukaryotic cells and are prominent in the long processes of the anisometric cells of the nervous system has led to their general acceptance as vectors guiding and stabilizing the complex shape of neurons. In addition to these tubules, there are 10-nm "intermediate" filaments (IF) (Figure 1-2) known as neurofilaments in neurons. These neurofilaments, the subject of this volume (but not this chapter), are considered to reside in stable or static zones of neuronal cytoplasm or in axons providing additional cytoskeletal properties. (For an excellent review of the distribution of neurofilaments in neurons, see reference 28.)

These filamentous structures of cells are suspended in material that appears flocculent in sections and is generally known as the cytoplasmic matrix, or ground substance. Recent evidence (to be detailed below) from electron microscopy indicates that the cytoplasmic

The authors would like to express appreciation to T. Deerinck, K. Anderson, G. Wray, J. Lindsey, and M. Beckerle for their work in these studies. We would also like to thank D. Taitano and J. Albersheim for assistance in typing this manuscript. This work was supported by grants from the National Institute of Neurological and Communicative Diseases and Stroke (NS14718) to MHE, from the Division of Research Resources (5P41RR00592) to KRP, from the Muscular Dystrophy Association to MHE and KRP, and from the National Multiple Sclerosis Society to MHE. MHE is an Alfred P. Sloan research fellow.

3

ground substance is highly structured and forms a three-dimensional lattice of slender strands, from 5 to 10 nm in diameter, referred to as microtrabeculae. One may consider the cytoplasm to be divided into two phases by these various formed elements: a water-rich fluid phase and a protein-rich polymerized phase (48). The water-rich phase is thought to include as solutes such small molecular metabolites as glucose and amino acids, diffusing O_2 and CO_2, and such important ions as K^+, Na^+, Ca^{2+}, and PO_4^{2-}. It is apparently devoid of larger species of molecules such as enzymes. Evidence from several model systems to be considered below indicates that the microtrabeculae are dynamic structures participating in transformations of cytoplasmic organization apparently promoting the rapid translocation of components within cells (intracellular transport). The microtrabecular lattice (MTL) takes on different forms in different cells (or portions of cells) and is not likely to be formed by the aggregation of a single macromolecule. It is, in other words, a heteropolymer. In keeping with this, several correlative arguments implicate a collection of dynamically interacting macromolecules possibly represented by the "slow component b" peptides described in axons by Black and Lasek (3, 4). We will consider some of the potential macromolecular components of this finely divided lattice later in this chapter. For now, however, it is sufficient to suggest that considering the microtrabecular lattice as a constitutive component of the cytoskeleton along with microtubules and neurofilaments may be inappropriate. This is because the microtrabecular components, by analogy, are likely to represent the dynamic subcellular musculature (cytomusculature) rather than the elements in cytoplasm providing a static cellular scaffolding, or "cytoskeleton." In developing or regenerating cells, the MTL exists as the system that probably decides where and in what orientation the cytoskeletal elements will reside in the cell.

Morphological Studies on the Fibrous Proteins

Microtubules

Most eukaryotic cells contain the now well-characterized 24-nm-diameter microtubule (Figure 1-1). These organelles are frequently found clustered around centrioles or near satellites of centrioles recognized as microtubule-nucleating centers (43). Microtubules are long, unbranching rods especially abundant in anisometric cells such as those that make up the nervous system (46, 50). Surrounding the microtubule there is a zone that excludes most common stains used in transmission electron microscopy (TEM). This "clear area" may be positively stained, however, after employing tannic acid as a mordant (Figure 1-2, inset) (60). The microtubule is known to be constructed of helically arranged alpha and beta subunits of tubulin that appear as 13 protofilaments in cross sections of the tubule (Figure 1-2, inset) (56). Microtubule subunits can be made to assemble and disassemble *in vitro*, and it is this attribute that has permitted their almost complete separation from other cellular components (66). Several varieties of microtubule-associated proteins (MAPS) copurify with tubulin through assembly-disassembly cycling techniques (39). The isolated MAP proteins are thought to play an important role in selectively associating microtubules with cell membranes and other subcellular organelles.

Microtubules are clearly involved in the maintenance of cell shape and in giving direction to the translocation of cytoplasmic components. By exposing cells to low temperatures or to the drug colchicine, anisometric cells or cell processes collapse concomitant with the disassembly of microtubules into their constituent subunits. The form of these cells or processes is reestablished upon tubule reassembly (59,61). Many of the very small cellular processes such as microvilli and lamelopodia lack microtubules but do contain actin-rich formed elements of the cytoplasmic ground substance (Figure 1-2). It does not appear that microtubules by themselves have contractile properties; instead, they appear to act as a frame for the attachment of closely associated macromolecular aggregates, like elements of the MTL, whose conformation changes in association with the translocation of granular or vesicular components of the cytoplasm.

Intermediate Filaments

Another ubiquitous fibrous protein is recognized in electron micrographs of most cells as the 10-nm-diameter "intermediate filaments" (IF) (Figure 1-2). (For recent reviews, see references 20 and 54.) Included in the general class of intermediate filaments are the 10-nm neurofilaments to be discussed in greater detail in later chapters of this volume. They are relatively stable structures that vary in chemical composition in different cell types. The role of intermediate filaments in intracytoplasmic motility or in providing for aniosometry of cells is less clear than that of microtubules; however, highly stabilized anisometric cells, such as neurons, contain very complex but highly reiterated networks of neurofilaments (28).

Neurofilaments, as well as analogous intermediate filaments in nonneuronal cells, have been shown to form cross-bridged complexes with microtubules both in neurons (13,16,35,42,53) and in fibroblasts (19). These mixed complexes have been noted in extended regions or cell processes where the microtubules and intermediate filaments run in parallel arrays. In most of the regions, and in particular within the cell body, aggregates of IFs are found. When observed

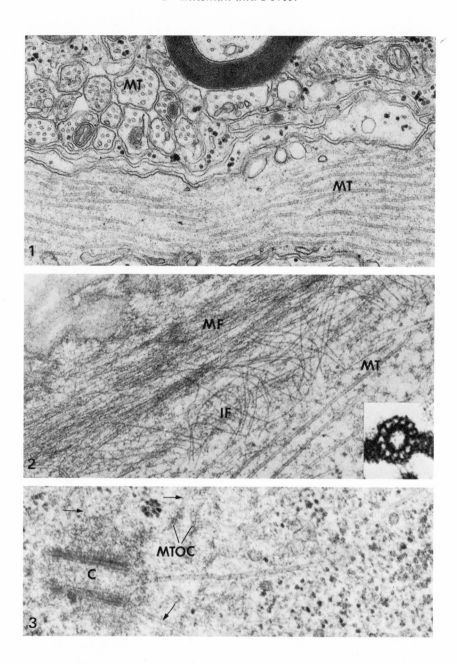

Figure 1-1. Micrograph of a thin section through the frog cerebellum. Many longitudinally oriented microtubules (MT) in a dendrite of a Purkinje cell are visible. Other smaller axonal or dendritic processes show microtubules in cross section. (Magnified × 35,000.)

Figure 1-2. Part of a thin section through the border region of a cultured WI 38 cell. An identifiable membrane (upper left) and microtubules in a stress fiber all are located

either by silver staining (67) or by immunofluorescence (19), these "neurofibrillar bundles" encapsulate the nucleus and extend into axonal processes. The neurofilament- and microtubule-rich zones of the perikaryon are distinct from zones of cytoplasm that contain granular endoplasmic reticulum or the Golgi apparatus (35). What role this compartmentalization plays and how stable it is with respect to the activities of the cell are questions as yet unanswered.

Microfilaments

Another filament type found in both muscle and nonmuscle cells is the 7 nm microfilament (MF) comprised primarily of the macro-molecule actin (Figure 1-2). Actin-myosin complexes along with other associated proteins form the highly specialized contractile element in muscle cells. There is little doubt that actin, in concert with myosin (25,44) or gelation and solation factors (for a review, see reference 51), can be utilized to generate contractile units such as stress fibers in nonmuscle cells. It is also evident that actin interacts with cell membranes and myosin in cells of differentiated tissues, such as lens photoreceptor cells or intestinal epithelia, to contribute to movement or to the generation and maintenance of aniosometric form (36,38,45). Actin is generally recognized in electron micrographs when in the form of filamentous actin (f-actin), but it is also known to associate with certain gelation factors to form complex, three-dimensional networks, so-called "actin gels" (9).

Is the Cytoplasmic Ground Substance Surrounding the Fibrous Proteins "The Microtrabecular Lattice"?

The distribution of these well-known components of the cytoskeleton is not random in cells. Microfilaments, for example, are confined to the peripheral zone of the cytoplasm. The so-called stress fiber is a differentiation of the cytoplasmic cortex and is never found deeper

in the cortical region of the cytoplast. The dense regions within the stress fiber are known to coincide with the distribution of alpha actinin. Below the stress fiber, there is a region rich in intermediate filaments (IF). These are mostly oriented normal to the long axis of the cell and normal to the orientation of the microfilaments (MF) (actin filaments) and the microtubules (MT). (Magnified ×35,000.) The small cross section of a microtubule (inset) showing protofilaments was derived from a paper from Tilney's laboratory (60).

Figure 1-3. Section through a centriole (C) at the pole of a mitotic spindle, found in a plasma cell located in the submucosa of the rat intestine. The continuity between the centriole-associated matrix and that which constitutes the general matrix of the spindle is visible. The image also includes a fairly good example of the connection between microtubules and a satellite microtubule-organizing center (MTOC). Small arrows point to MT insertions. (Magnified ×35,000.)

in the cytoplast (Figure 1-2). Furthermore, stress fibers, as observed in cultured cells, do not occupy fixed positions. Rather, they disassemble and reassemble, and change their orientation, as the cell moves over the substrate or alters its shape. Arrays of actin filaments can be identified by heavy meromyosin decoration and are a characteristic feature of lamellopodia and intestinal microvilli.

Microtubules are likewise nonrandom in distribution and orientation, especially in differentiated cells. Ordinarily, the proximal ends of the tubules are inserted into one of several dense bodies (tubule-initiating sites) that comprise, with centrioles, the cell center (centrosome or centrosphere) (Figure 1-3). In tissue cells examined *in situ*, they are seen to adopt patterns that may repeat more or less precisely from cell to cell (7). This impressive evidence of organization is lost by cells in culture, presumably because they are proliferating and/or are not contact inhibited and differentiated. Neoplastic cells, in contrast with normal cells, are distinguished for showing more random arrangements of microtubules.

The so-called intermediate filaments are less discriminating in their distribution (Figure 1-2). This variation repeats, in a sense, the variation shown by the composition of these filaments and the roles they apparently perform. In general, it seems that they are designed and distributed to resist stresses or to provide physical stability to the cytoplasmic heteropolymers of which they are part. It would then be a mistake, on the strength of these comments, to conclude that these 10-nm filaments are exceptional in not displaying any organization. In fact, they are similar to the other polymers in showing order, both in distribution and orientation in cells of the same type. They differ in appearing to occupy almost any positions, central or peripheral, in the cytoplast in cells of different types.

Of the many interesting questions that are asked of the cytoskeletal components and that grow out of the above observations, perhaps none is more important than what controls or guides their assembly. Microtubules, for example, appear to be initiated at microtubule-organizing centers (MTOCs, dense bodies) that are dispersed in some pattern relative to the centrioles (Figure 1-3). Structurally, the centrioles are continuous with the dense bodies through a condensed version of the MTL. Similarly, the latter is a structural extension of the centrosphere. There are no separating membranes or spaces. Does it follow then that microtubule assembly takes place within the trabeculae of the MTL, or is it dependent on available monomeric tubulin in the water-rich phase of the intertrabecular spaces. The evidence with which to select among such alternatives is meager. There is some suggestion from immunoperoxidase preparations (involving the use of specific antitubulin antibodies) that tubulin is available in narrow linear zones within the lattice and that assembly follows these zones (10). Then also, as stated earlier, there is no evidence of protein in the water-rich phase. One inclines to the view that tubulin is available nonrandomly and that its synthesis takes place in predetermined lattice-related sites where appropriate messenger RNA and polysomes reside.

The general properties of cells of a single type include their uniform size, their organization, their form, and their capacity to recognize the loss of a part and to regenerate it; all of these describe the existence of a structural unit, a continuum, in which the various organelles and systems are nonrandomly dispersed and nonrandomly assembled. In recent years, the search for the physical expression of such a continuum has been facilitated by high-voltage microscopy and the application of new techniques of specimen preparation. Thus, whole cells, cultured *in vitro*, have been examined for their three-dimensional structures. What has emerged is an image depicting the presence of a three-dimensional meshwork, or lattice, of fine strands. This comprises the cytoplasmic matrix or ground substance heretofore regarded as structureless (Figures 1-4 and 1-5). It has been identified in all cells examined; it fills the space between the nuclear envelope and the inner surface of the plasma membrane. It is constructed of fine trabeculae, microtrabeculae, which are continuous with one another and the surfaces of all the better-known formed structures of the cytoplasm, with the possible exception of mitochondria. Its major features are depicted in the stereo images shown in Figure 1-5 and in the drawing (model) included as Figure 1-6. The following comments refer to what is shown in the model.

The presence of this structure effectively divides the matrix into two phases: one protein-rich and the other water-rich. The latter is represented by the intertrabecular spaces that are apparently without content. The lattice (Figure 1-6) supports or contains in its meshes the microtubules as well as microfilaments and intermediate filaments (not shown). Free ribosomes or polysomes likewise occupy positions within the lattice and are thought to contribute the protein they synthesize directly to the lattice (18) (Figure 1-6). If messenger RNA is nonrandomly distributed, as it would seem to be, then different protein species may be nonrandomly dispersed in the lattice and in the cell (as just discussed). It is supposed that by this device such products of differentiation as myofibrils are positioned within the cell where the constituent proteins are synthesized. There is good evidence that the lattice is dynamic in the sense that the structural elements, the trabeculae, are constantly shortening and elongating. Clearly, the structure is one that would be expected of a viscoelastic gel. It is known to decay structurally when the cell is maintained over

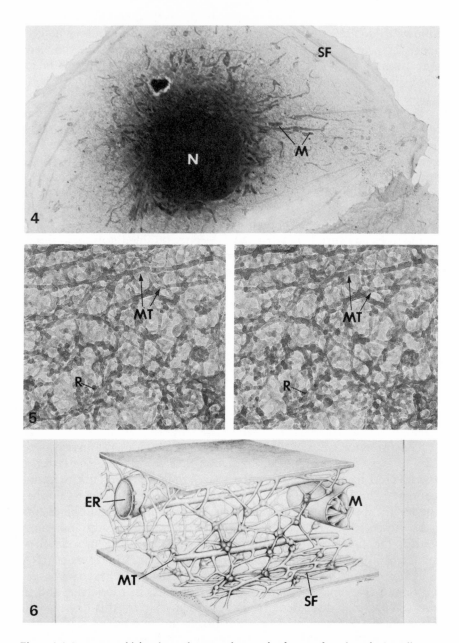

Figure 1-4. Low-power high-voltage electron micrograph of most of a cultured BSC cell (kidney of African green monkey). It shows stress fibers and mitochondria and a reasonably unstructured (at this magnification) cytoplasmic matrix. It serves as background for the pair of stereos in Figure 1-5. (Magnified × 1,800.)

2 to 3 hours at 4°C. But the components do not go into solution. In cells returned to 37°C after chilling, the lattice is quickly (within 5 minutes) rebuilt. The water-rich phase, by its presence, can accommodate about 60% of the cell's water. The volume of this phase is easily affected by exposing a cell to either hyper- or hypotonic environments, and this without destroying the MTL or the cell's viability. Meaningful data on the composition of the lattice is difficult to obtain. Detergent extracts of cells, examined by two-dimensional electrophoresis, contain over 400 different polypeptides. Some of these derive from membranes and ribosomes as well as the content of the endoplasmic reticulum (ER) and the Golgi complex. It is most accurate to say that such extracts contain everything except the proteins of the cytoskeleton, and even some of these are present. Thus, the analyses are scarcely descriptive of the lattice. Even if one could isolate the lattice for examination of its content, one would be homogenizing a structure that probably varies in composition from one part to another. In this discussion, the reader should be reminded that evidence of compartmentalization in the distribution of all proteins is beginning to accumulate.

MTL: Artifact versus Reality

When considering extremely fine structured details within cells such as the microtrabecular lattice, one must be concerned about the possibility of artifact. Could this fine lattice have been created artifactually by fixation or significantly modified from its real state in a living cell? These questions have been vigorously investigated in recent years through the use of several different systems (16,47,71). The same lattice structure has been observed within cultured cells after fixation by aldehydes, osium tetroxide, or rapid freezing followed by critical point drying or freeze-drying. Regardless of the preparative procedure, aspects of this cross-linked structure are seen as essentially the same with only small variations in the dimensions of the elements. When cells were examined in various tissues by using selective staining or a soluble compound such as PEG for embedding (16,22,69,72), it was found that the microtrabecular lattice is universally present and can take the form of periodic cross-bridges between linearly oriented

Figure 1-5. Stereo pair of high-voltage micrographs obtained from the margin of a thinly spread BSC cell. It shows microtubules (MT), ribosomes (R), and microtrabeculae (the MTL). (Magnified ×80,000.)

Figure 1-6. Model of the lattice with "contained" microtubules, ribosomes, and cisternae of the ER. Stress fibers are depicted as local differentiations of the MTL. (See text for more detail.)

fibrous proteins (this is especially true in axons and axonemes). The consistent observations made through varied preparative techniques and the periodic nature of the microtrabecular cross-bridges in axons argue strongly for the existence of a correlate to these structures in living cells.

Dynamic Properties of the Cytoskeleton: Participation of Fibrous Proteins in Dynamic Events

The dynamic properties of microtubules and intermediate filaments have been investigated with *in vitro* reassembly systems (58,66). Cyclic assembly of tubulin has been considered as a possible mechanism fo some forms of intracellular motility *in vitro* (34). Although microtubules appear far less stable than neurofilaments in extraction buffers (37), there is at present no direct evidence indicating tubule reassembly as a mechanism for rapid forms of intracytoplasmic motility. Some other observations can be interpreted as supporting the concept of stable microtubules, at least within axons. Very long microtubules (over 500 μm in length) have been observed in axons. Labeled tubulin subunits move very slowly (1 to 2 mm per day) down the axon along with the neurofilament proteins (24,27).

Numerous reports correlate the effects of cytoskeleton-disrupting procedures on intracytoplasmic motility in neurons. Treatment of axons with colchicine or vincristine, for example, inhibits microtubule formation and according to most reports also disrupts axonal transport. (For a recent review, see reference 21.) Likewise, other procedures that disrupt microtubules, such as lowered temperatures and elevated Ca^{2+} concentrations, also impair transport (5,40). Neurofilaments in axons do not appear to be affected by these procedures. Also, neurofilament subunits do not show up rapidly in extraction buffers that allow leeching out of axonal components close to their equilibrium concentration (37). The effect of microtubule-disruptive procedures on transport provides sufficient evidence for most to agree that rapid axonal transport is based on a microtubule-related motility system. Data on the actual rate of tubulin movement, however, indicate that the rapidly moving components move relative to microtubules (4,24,68). This relative motion sets the interfacial material (the MTL) in a key position to regulate and provide the motive force for rapid axoplasmic transport. Cross-bridges have been noted between the rapidly moving membrane-bounded components and microtubules or neurofilaments. (See Chapter 4.) A dynamic component of MTL morphology (functional changes) has been noted as well in chromatophores. (See below.) It is noteworthy in this regard that the cross-bridges between microtubules and between neurofilaments are differentially susceptible to detergent extraction, the latter being much more stable (65).

Thus, when rapid intracellular motility is being considered, the dynamic aspects of microtubules and neurofilaments are probably not so important as the dynamic properties of the MTL; rather, the fibrous proteins can be considered stators against which force-generating macromolecules push or pull. However, the dynamic (elastic) properties of these two types of fibrous proteins may be important in contributing to static or dynamic aspects of cell shape or development. In this case, the dynamic properties of microtubules probably provide plasticity, while the stable properties of neurofilaments impart elastic rigidity. (For an excellent discussion of this subject, see reference 28.)

The Neuronal Cytoskeleton

At the beginning of this chapter, we noted the prominent fibrillar staining surrounding the nucleus in silver-stained neurohistological preparations. With appropriately accurate hindsight, it is not surprising that the neurofilament networks correspond with the fibrillar cytoskeleton. The endomembrane system of the neuron is adjacent to, and somewhat coincidental with, the concentrations of fibrillar material surrounding the nucleus and extending into the axons and dendrites. This system is composed primarily of the organelles functioning as the sites for protein synthesis, glycosylation, and packaging of materials for export. The primary synthesis of proteins for export occurs at the level of the ribsome-laden or rough endoplasmic reticulum (RER). The endomembrane and fibrous protein systems thus associate in the soma and maintain a close proximity in the long and highly specialized neuronal processes.

The actin-containing, microfilament-rich regions of the neuron appear to be distinct from the microtubule-neurofilament-rich zones. Microfilaments are found more peripherally located in axonal processes, often in close association with the plasma membrane (26). Numerous morphological specializations of the neuronal plasma membrane have been noted, many with associated microfilamentlike structures penetrating deeper into the cytoplasm, there presumably associating with the "core" microtubule/neurofilament cytoskeleton. For example, the postsynaptic density of the synapse has associated actinlike filamentous material (23). The axolemma at initial segments of axons and at nodes of Ranvier is specialized with submembranous material (41) that is also filamentous and connected to the core cytoskeleton (12,16). Actin has also been localized to dendritic spines (31) and, with the MTL and cytoskeletal proteins, may play a role in the morphological changes noted for these postsynaptic specializations.

Thus, the "core" neuronal cytoskeleton is implicated in the routing of synthesized components from the cell body into and down neuronal processes. It is also cast in the role of stator for rapid neuroplasmic transport processes. The "core" cytoskeleton further acts as a substratum for the attachment of microfilament systems impinging upon the plasma membrane. Local specialization of this microfilament system provides for regional specializations of the plasma membrane and in concert with other cortical macromolecules participates in anchoring transmembrane proteins (such as neurotransmitter receptors) in place. Embedded within the core cytoskeleton of axons are several membrane systems (some motile, some not) that are extensions of the endomembrane system (ER and Golgi) noted earlier. The interactions between these axonal membrane systems and the cytoskeleton are currently a topic of great interest bearing on intracytoplasmic motility mechanisms. Some recent findings on the axonal membrane systems are related in the next section.

Membrane Systems of the Axon

A number of selective staining techniques have been used on axons and have revealed a membrane system referred to as "the axoplasmic reticulum" (Figure 1-7). Components of this system have been described by many workers (11,62). Figure 1-7 illustrates the form and distribution of this reticulum within the axoplasmic space of a myelinated axon. The function of this system is unknown. It has been suggested that some materials may move through this cisternal system or that this cisternal system itself moves down the axon. (For a review, see reference 21.) It has been known for some time that rapidly transported macromolecules move in association with membranous structures. In an important recent study, Tsukita and Ishikawa (63) showed that a noninvasive focal cold blockade of axonal transport, achieved by simply gluing an aluminum block (cooled to 4°C) over the saphenous nerve (Figure 1-8), results in materials moving in the orthograde direction building up on the proximal side of the cold block while retrogradely moving components on the proximal side continue to move away from the site. Conversely, materials moving retrogradely on the distal side of the block accumulate, while orthogradely moving components on the distal side move on. Thus, one can examine by electron microscopy the morphology of retrograde and orthograde compartments after a suitable incubation. A similar study was recently conducted by Smith (55), who used a mechanical block on single fibers.

Micrographs taken from tissue on the proximal side of the blockade consistently demonstrate an accumulation of vesicular or vesiculo-tubular materials. These types of discrete membrane-bounded compartments accumulate where orthogradely moving components would be expected to accumulate selectively. Figure 1-9 represents the orthograde transport compartments or "vectors" as defined by laboratories conducting such accumulation experiments, including our own. Since these accumulating compartments did not resemble the axoplasmic reticulum of Figure 1-7 and we did not notice its accumulation against either distal or proximal sides of the cold block, we became curious about its mobility.

We have taken advantage of the above-described approach and designed a double cold block system (Figure 1-10) to test whether or not the axoplasmic reticulum moved rapidly (14,15). The double cold block apparatus enables us to hold a small central region of the axon at a constant temperature of 37°C. After 3 to 4 hours, any elements moving faster than 25 mm per day should move to either the orthograde dam or the retrograde dam. Organelles remaining in the center of the warm zone should not be rapidly transported components. In these experiments, we found that elements of the axoplasmic reticulum are the major component remaining in such central regions (Figure 1-11) (14,15). Even at nodes of Ranvier, where one normally find more orthograde or retrograde vectors than in internodal segments, elements of the axoplasmic reticulum are the primary component found in sections from the middle zone of the double cold block preparation. In the region of the proximal dam where orthograde vectors accumulate, one can see clearly that the axoplasmic reticulum winds its way through the discrete orthograde vectors without apparent fusions (Figure 1-12). These observations fuel our speculation that the axoplasmic reticulum is a distinct membrane system within the axon, separate from the motile membranous compartments that carry rapidly transported macromolecules.

Limitations of Present Information

The discussion of the microtrabecular lattice morphology in an earlier section of this chapter was derived mainly from the results obtained from a variety of thinly spread cultured cells. The cultured cell lends itself well to examination by many techniques in which embedment and epoxy resins are avoided or unnecessary. Unfortunately, aside from the examination of cultured neurons, most observations on the cytoplasm of neurons must be made in sections of cells from tissues embedded in epoxy resins. When one examines conventional thin sections of neurons (or for that matter other cells embedded in epoxy

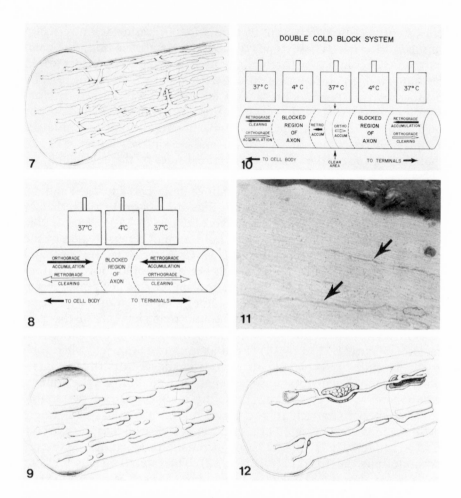

Figure 1-7. An artist's representation of the membrane systems of the axon drawn approximately to scale in the same cytoplasmic space: The axoplasmic reticulum.

Figure 1-8. Diagram illustrating the design of the single cold block experiments. The central area is blocked, while orthogradely moving materials accumulate proximally and retrogradely moving materials accumulate distally.

Figure 1-9. An artist's representation of the membrane systems of the axom drawn approximately to scale in the same cytoplasmic space: Orthograde transport compartments or vectors.

Figure 1-10. Diagram illustrating the design of the double cold block experiments. The cylinder represents axons within the saphenous nerve, while the squares represent the selectively warmed and chilled blocks of aluminum applied to the skin just above the nerve. Beneath the two 4°C blocks, fast axonal transport is arrested; under the 37°C blocks, it continues in a normal fashion. After 3 to 4 hours, transported material moving faster than 25 mm per day in either direction will have cleared from the region under the central warm block and no new material will have entered.

Figure 1-11. In the central warm region, between the two cold blocks, fast transported

resins), looking for an equivalent of the microtrabecular lattice, difficulties are immediately encountered. The difficulties evolve from the fact that the electron-scattering properties of the resin coincide with the scattering properties of many fine structural components. Traditionally, one circumvents this problem by the addition of heavy-metal stains to enhance the electron scattering associated with various subcellular structures. Hence, mordants such as tannic acid, mentioned above in association with microtubules, enhance the staining of tubulin by uranyl acetate and osmic acid. In the case of the cultured cell viewed after critical-point drying, much of the structure visualized in the electron microscope is visible as a result of the electron scattering inherent in the macromolecular aggregates themselves and the relative absence of scattering matter in the intertrabecular spaces. Thus, techniques must be used that selectively enhance the electron scattering of the finest subcellular structures significantly above the electron scattering of the epoxy resin, or techniques must be used that allow for the removal of embedding material such as polyethylene glycol from the sections.

Another factor contributing greatly to the recognition of a correlate for the microtrabecular lattice in sections of cells from tissues is three-dimensional visualization. In conventional two-dimensional electron micrographs derived from thin section, many fine structures appear as wispy, disconnected elements because of the misrepresentation of superpositioned structures recorded in two-dimensional images. When thick or even thin sections are viewed with the aid of stereopair electron micrographs, obtained through the use of modern goniometric accessories for the electron microscope, considerable information concerning the distribution of cross-linking elements may be obtained. For example, view Figure 1-13 as a single two-dimensional image and then view it again using a stereo viewer. You will notice that many of the elements cross-linking microtubules and neurofilaments in this pair of micrographs, when viewed nonstereoscopically, appear to end blindly and might be interpreted as not being noticeably tied together in a two-dimensional image. In sections of this sort, stained *en bloc* with uranyl acetate, the interconnecting wispy strands of trabeculae appear to radiate into the ground substance at approximately right angles to the microtubules and neurofilaments (Figure 1-13). These cross-bridges often connect microtubules with

elements that have presumably cleared from the axoplasm. In this 1-μm-thick section viewed with the high-voltage electron microscope (HVEM), the only membranous element remaining in the axoplasm is the axoplasmic reticulum (arrows). (Magnified $\times 24,000$.)

Figure 1-12. An artist's representation of the membrane systems of the axom drawn approximately to scale in the same cytoplasmic space: Association between the axoplasmic reticulum and both orthograde (lower) and retrograde (upper) vectors.

microtubules, neurofilaments with neurofilaments, as well as microtubules with neurofilaments. Not only do these cross-bridges link the major fibrous proteins of the axon together, they also link these fibrous proteins with membranous organelles such as multivesicular bodies or smooth membranous cisternae resident in the axon or other regions of the neuron (15,16).

New Techniques

As noted above, one reason that the microtrabecular lattice is not easily recognized in conventional thin sections is that the electron scattering of epoxy resins coincides with that of some fine structural details. Enhancement of epoxy-section images with heavy-metal stains can compensate somewhat for this effect; however, this is often quite incomplete. Three-dimensional visualization of stained sections can allow for an increased appreciation of the interactions between such fine details, but, if these structures are not sufficiently electron dense, they are not visualized clearly in the electron micrographs.

Several new techniques have emerged that circumvent some of the problems inherent in epoxy embedding. As mentioned above, we have used one of these, polyethylene glycol embedding (PEG), in order to observe thick or thin sections from which the embedding matrix has been removed (16,22,69,72). PEG is a water-soluble wax that may be removed from ultrathin sections by gentle washing in warm water. Figure 1-14 is an image from an axon sectioned after PEG embedding.

The fibrous nature of the axoplasm is easily appreciated. At high magnifications and in stereomicrographs, the cross-bridges are readily visualized and closely resemble the microtrabecular cross-bridges of the critical-point-dried or freeze-dried cultured cells described earlier. Another method of viewing the PEG-embedded material is scanning EM (SEM). Figure 1-15 is a scanning electron micrograph of the block face from which sections had been cut. In these preparations, the PEG is washed from the block and the piece of axon dehydrated, critical-point-dried and coated with a thin layer of gold to be viewed by SEM. Although visibility of the fibrous proteins and the cross-bridging elements is enhanced with these PEG techniques, it is more difficult to recognize the membranous inclusions of the axon. To examine the interaction among such membrane systems and the fibrous proteins, more conventional techniques are used.

Cross-Bridges and the Directionality of Transport Vectors

We have described the cross-linking structures of the cytoskeleton (cytomusculature), the various ways these are represented in axons, and the membrane systems of the axon. These elements appear to interact dynamically, as evidenced by the way in which the lattice is specialized where it interacts with the motile membranous components of the axon. Neurofilaments and microtubules are cross-linked in a relatively periodic manner by wispy elements (16,65). In our studies, we found that small cisternae often appeared asymetrically connected to microtubules or neurofilaments by these wispy elements, as is seen with the small vesicle of Figure 1-13.

By using the single focal cold block method for stopping transport (Figure 1-8), we were able to look, not in the area of accumulation, but in the area where orthograde vectors would be moving toward the accumulation against the dam. We reasoned that, if given a sufficient length of time, these orthograde vectors, in route, would be isolated from the retrogradely moving vectors, since the retrograde vectors would presumably continue to move on toward the cell body. Retrograde vectors have been shown to be generally multivesicular bodies, multilamellar bodies, and clear and dense lumed cisternae (15,30,55,63). By maintaining the proximo-distal orientation of the axon all the way to the electron micrograph, we were able to determine that asymmetry of cross-bridges and the organization of microtubules and neurofilaments relates to vector direction. There is a predominance of cross-bridges on the leading ends of vectors and an absence of cross-bridges (as well as large spaces between filaments) on the trailing ends of vectors. We suspect that this asymmetry has something to do with the actual movement of vectors during rapid axonal transport. For example, the cisterna in Figure 1-16 is presumably moving from right to left in the axon. Figure 1-17 summarizes our observations on the asymmetry of cross-bridges on both the orthograde vectors. The same type of observation has been made for retrograde vectors where the polarity is similar (14).

Although this new data on axoplasmic vector direction implicates the involvement of cross-bridging elements in intracellular motility (in this case, rapid axonal transport), more direct evidence comes from work with cultured cells. The cultured cell systems provide a somewhat more manipulable and observable preparation.

Dynamic Properties of the MTL in the Model Systems

The chromatophores of fishes are excellent objects for the study of intracellular transport. Derived as they are from the neural crest, they may be regarded as nerve cell related. Their pigment granules can be observed in the light microscope; and the responses of these cells to such exogenous agents as epinephrine, cAMP, caffeine, nocodazole, and metabolic inhibitors can be followed in living cells. We (together with colleagues) have studied these and other responses especially in

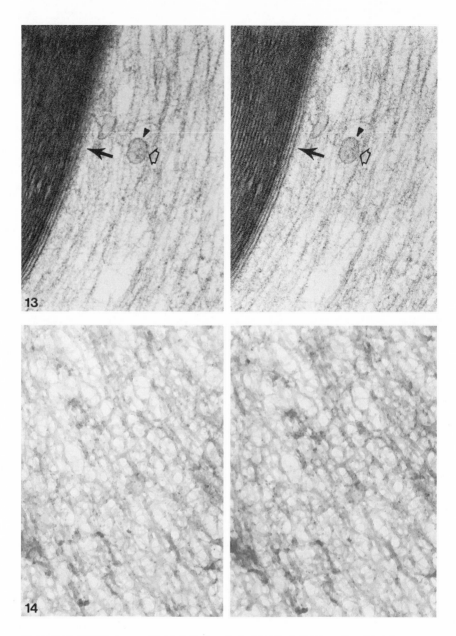

Figure 1-13. HVEM stereo pair micrographs of a longitudinal, thick section through a myelinated peripheral nerve axon. Linkages to the plasma membrane (solid arrows) and cisternae of axonal SER (open arrows) are visible. Many small vesicular fragments of the SER (orthograde vectors) are larger in diameter than the average distance between adjacent neurofilaments and microtubules. The cross-bridging connections are often disconnected on one end of such cisternae (arrowhead). (Magnified ×75,000.)

Figure 1-14. Stereo pair electron micrographs of a section cut from PEG-embedded peripheral nerve. At high magnification, the trabecular cross-bridging connections between microtubules and neurofilaments, similar to those found in cultured cells, are viewed without an embedding matrix. (Magnified ×60,000).

erythrophores isolated from the scales of *Holocentrus ascensionis* (Osbeck) and have correlated our observations with the fine structure as observed by high-voltage electron microscopy of whole cells (1,8, 32).

Fortunately, erythrophores can be maintained in culture, where they aggregate and disperse their pigment as in the scale. In aggregation, the pigment moves at a uniform velocity of about 10 μm per second and covers the distance from cell periphery to center in 2 to 3 seconds. Dispersion is a different phenomenon; in this, the motion is saltatory and is completed in roughly twice the time required for aggregation. If during either of these motions the cells are quickly fixed and processed for electron microscopy, the motion-associated structural changes can be observed. In the dispersed state, the pigment granules are supported (or contained) in a meshwork of fine strands, 5 to 10 nm in diameter. These interconnect the pigment granules and the microtubules and constitute an irregular three-dimensional lattice resembling in all respects the microtrabecular lattice (MTL) of other cell types. When the cell is stimulated with epinephrine to aggregate its pigment, or with high K^+, the individual trabeculae thicken and shorten and the whole meshwork, with pigment, contracts into a compact, spheroid mass around the cell center. In dispersion, the lattice expands again into its ellipsoid form as the microtrabeculae lengthen; it appears to be restructured under the vectorial guidance of the radially organized framework of microtubules. This restructuring and the associated redispersion of the pigment is ATP dependent. In cells deprived of ATP, the pigment gradually aggregates as it does when the cell dies (32). It appears that elastic energy stored in the lattice during dispersion is released in aggregation (32) and that Ca^{2+} initiates these events (33). There is, in fact, a complex system of smooth ER vesicles (SER) associated with the lattice and microtubules that may control Ca^{2+} concentration and distribution much as the sarcoplasmic reticulum does in muscle (49). It is evident from the available data that the microtrabeculae, through conformational changes, provide the motive force for both aggregation and dispersion and that the microtubules give guidance to the motion.

These three components of the erythrophore plus the centrosphere comprise the cytoplast. We have recently observed that the individual pigment granules occupy fixed positions (FP) within the cytoplast (49). Even though a dramatic shift occurs in the shape of the cytoplast associated with aggregation (from ellipsoid to spheroid), the granules remain in a FP relative to the whole pigment mass. This is mentioned to emphasize that cytoplast organization at this level of fine structure is apparently a property of these and probably all cells.

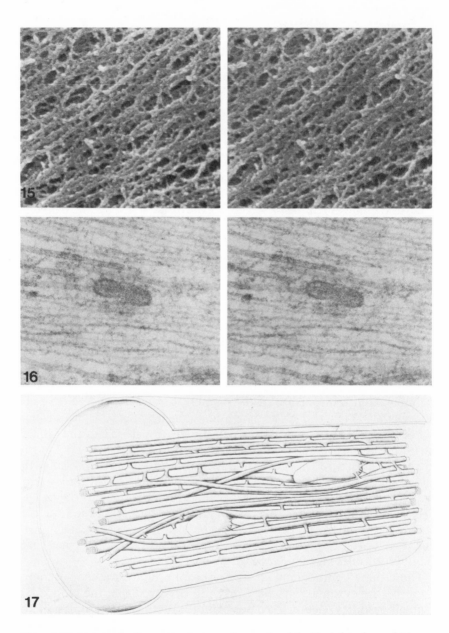

Figure 1-15. Stereo pair of scanning electron micrographs of the block face from which PEG sections have been cut. Note the numerous lateral connections between neurofilaments and microtubules. (Magnified ×50,000.)

Figure 1-16. An orthograde vector presumably moving from left to right with crossbridges on the leading end and a clear area behind. (Magnified ×100,000.)

Figure 1-17. An artist's representation of the asymmetry of cross-briding and of differences in microtubule and neurofilament alignment (organization) as the orthograde vectors illustrated move from left to right.

Recently, in our attempts to define better the characteristics of the energy-transducing unit, we have utilized procedures for micro-injection and permeabilization of these cells that permit us to introduce into the environment of the MTL such dynein-ATPase inhibitors as EHNA and vanadate (2,57). Both agents have been observed to stop pigment motion, and the vanadate effect is reversible. This suggests that the microtrabeculae in this system may be equated with the dynein complexes of cilia.

Conclusions

In closing, the discussions above illustrate that axoplasm shows in its fine structure several of the same features as shown by the cytoplast of erythrophores; the MTL is, in fact, common to all cells. These can be seen in replicas derived from frozen-fractured and etched axoplasm as well as from thin sections (16,52,65) and are thus not artifactually derived. Concepts of function in these two systems as well as in cilia and the mitotic spindle are being discussed and examined for similarities. In uncovering structural similarities, we will undoubtedly uncover differences between the neuron and the erythrophore. Clearly, the cytoskeletal/muscular system we are examining is dynamic. Perhaps more importantly, however, our concept of its properties is growing rapidly with the acquisition of new knowledge.

REFERENCES

1. Beckerle, M. C., H. R. Byers, K. Fujiwara and K. R. Porter. 1979. Indirect immuno-fluorescent stereo high voltage electron microscopy evidence for microtubule associated migration of pigment granules in erythrophores. *J. Cell Biol. 83*:352 (abs.)

2. Beckerle, M. C., and K. R. Porter. 1982. Inhibitors of dynein activity block intra-cellular transport in erythrophores. *Nature 295*:701-703.

3. Black, M. M., and R. J. Lasek. 1979. Axonal transport of actin: Slow component B is the principal source of actin for the axon. *Brain Res. 171*:401-413.

4. Black, M. M., and R. J. Lasek. 1980. Slow components of axonal transport: Two cytoskeletal networks. *J. Cell Biol. 86*:616-623.

5. Brimijoin, S., J. Olsen, and R. Rosenson. 1979. Comparison of the temperature-dependence of rapid axonal transport and microtubules in nerves of the rabbit and bullfrog. *J. Physiol. 287*:303-314.

6. Brinkley, B. R., S. M. Cox, and S. H. Fistel, 1981. Organizing centers of cell processes. *Neurosciences Res. Prog. Bull. 19*:108-124.

7. Byers, H. R., K. Fujiwara and K. R. Porter. 1982. Visualization of microtubules of cells *in situ* by indirect immofluorescence. *Proc. Natl. Acad. Sci. USA 77*:6657-6661.

8. Byers, H. R., and K. R. Porter. 1977. Transformations in the structure of the cytoplasmic ground substance in erythrophores during pigment aggregation and dispersion. *J. Cell Biol. 75*:451-558.

9. Condeelis, J. S. 1981. Reciprocal interactions between the actin lattice and cell membrane. *Neurosciences Res. Prog. Bull 19*:83-99.

10. DeMay, J., J. J. Wolosewick, M. DeBrabander, G. Gevens, M. Joniau, and K. R. Porter. 1980. Tubulin localization in whole glutaraldehyde fixed cells, viewed with stereo electron microscopy. *In*: Cell Movement and Neoplasia. M. DeBrabander, M. Mareel, and L. DeRidder, editors. Pergamon Press, Oxford and New York, pp. 21-28.

11. Droz, B., A. Rambourg, and H. L. Koenig. 1975. The smooth endoplasmic reticulum: Structure and role in the renewal of axonal membrane and synaptic vesicles by fast axonal transport. *Brain Res. 93*:1-13.

12. Ellisman, M. H. 1977. High voltage electron microscopy of cortical specializations associated with membranes at nodes of Ranvier. *J. Cell Biol. 75*:108 (abs.).

13. Ellisman, M. H. 1981. Beyond neurofilaments and microtubules. *Neurosciences Res. Prog. Bull. 19*:43-58.

14. Ellisman, M. H., and J. D. Lindsey. 1981. The axonal reticulum in myelinated axons is not rapidly transported and an asymmetry of cross-bridges exists in cisternae that are transported. *J. Cell Biol. 91*:91 (abs.).

15. Ellisman, M. H., and J. D. Lindsey. 1983. The axoplasmic reticulum within myelinated axons is not rapidly transported. *J. Neurocytol.*, in press.

16. Ellisman, M. H., and K. R. Porter. 1980. The microtrabecular structure of the axoplasmic matrix: Visualization of cross-linking structures and their distribution. *J. Cell Biol. 87*:464-479.

17. Fifkova, E., and A. Van Harreveld. 1977. Long-lasting morphological changes in dendritic spines of dentate granular cells following stimulation of the entorhinal area. *J. Neurocytol. 6*:211-230.

18. Fulton, A. B., K. M. Wan, and S. Penman. 1980. The spatial distribution of the polyribosomes in 3T3 cells and the associated assembly of proteins into the skeletal framework. *Cell 20*:849-857.

19. Goldman, R. D., B. Chojnacki, A. E. Goldman, J. Starger, P. Steinert, J. Talian, M. Whitman, and R. Zackroff. 1981. Aspects of the cytoskeleton and cytomusculature of nonmuscle cells. *Neurosciences Res. Prog. Bull. 19*:59-82.

20. Goldman, R. D., A. Milsted, J. A. Schloss, J. Starger, and M.-J. Yerna. 1979. Cytoplasmic fibers in mammalian cells: Cytoskeletal and contractile elements. *Ann. Rev. Physiol. 41*:703-722.

21. Grafstein, B., and D. S. Forman. 1980. Intracellular transport in neurons. *Physiol. Rev. 60*:1167-1283.

22. Guatelli, J. C., K. R. Porter, K. L. Anderson, and D. P. Boggs. 1982. Ultrasctructure of the cytoplasmic and nuclear matrices on human lymphocytes observed using high voltage electron microscopy of embedment-free sections. *Biol. Cell 43*:69-80.

23. Gulley, R. L., and T. S. Reese. 1982. Cytoskeletal organization at the postsynaptic complex. *J. Cell Biol. 91*:298-302.

24. Hoffman, P. N., and R. J. Lasek. 1975. The slow component of axonal transport: Identification of major structural polypeptides of the axon and their generality among mammalian neurons. *J. Cell Biol. 66*:351-366.

25. Kron, E. D. 1978. Biochemistry of actomyosin-dependent cell motility (a review). *Proc. Natl. Acad. Sci. USA 75*:588-599.

26. Kuczmarski, E. R., and J. L. Rosenbaum. 1979. Studies on the organization and localization of actin and myosin in neurons. *J. Cell Biol. 80*:356-371.

27. Lasek, R. J. 1980. Axonal transport: A dynamic view of neuronal structures: *Trends Neurosci. 3*:87-91.

28. Lasek, R. J. 1981. The dynamic ordering of neuronal cytoskeletons. *Neurosciences Res. Prog. Bull. 19*:7-32.

29. Lasek, R. J., and M. L. Shelanski. 1981. Cytoskeletons and the architecture of nervous systems. *Neurosciences Res. Prog. Bull. 19*:3-153.

30. La Vail, J. H., S. Rapisardi, and I. K. Sugino. 1980. Evidence against the smooth endoplasmic reticulum as a continuous channel for the retrograde axonal transport of horseradish peroxidase. *Brain Res. 191*:3-20.

31. LeBeux, Y. J., and J. Willemot. 1975. An ultrastructural study of the neurofilaments in rat brain by means of E-PTA staining and heavy meromyosin labeling: II. The synapses. *Cell Tissue Res. 160*:37-68.

32. Luby, K. J., and K. R. Porter. 1980. The control of pigment migration in isolated erythorophores of *Holocentrus ascensionis* (Osbeck): I. Energy requirements. *Cell 21*:13-23.

33. Luby-Phelps, K. J., and K. R. Porter. 1982. The control of pigment migration in isolated erythropores of *Holocentrus ascensionis*: II. The role of calcium. *Cell, 29*:441-450.

34. Margolis, R. L., and L. Wilson. 1979. Regulation of the microtubule steady state *in vitro* by ATP. *Cell 18*:673-679.

35. Metuzals, J., and W. E. Mushynski. 1974. Electron microscope and experimental investigations of the neurofilamentous network in Deiter's neurons: Relationship with the cell surface and nuclear pores. *J. Cell Biol. 61*:701-722.

36. Mooseker, M. S., and L. G. Tilney. 1975. Organization of an actin filament-membrane complex: Filament polarity and membrane attachment in the microvilli of intestinal epithelial cells. *J. Cell Biol. 67*:725-743.

37. Morris, J., and R. J. Lasek. 1979. Differential solubilies of cytoskeletal proteins in squid axoplasm. *Biol. Bull. 157(2)*:384.

38. Mukherjee, T. M., and L. A. Staehelin. 1971. The fine-structural organization of the brush border of intestinal epithelial cells. *J. Cell Sci. 8*:573-599.

39. Murphy, D. B., and G. G. Borisy. 1975. Association of high-molecular-weight proteins with microtubules and their role in microtubule assembly *in vitro*. *Proc. Natl. Acad. Sci. USA 72*:2696-2700.

40. Ochs, S., R. M. Worth, and S.-Y. Chan. Calcium requirement for axoplasmic transport in mammalian nerve. *Nature 270*:748-750.

41. Peters, A. 1966. The node of Ranvier in the central nervous system, *Quart. J. Exp. Physiol. 51*:229-239.

42. Peters, A., S. L. Palay, H. deF. Webster. 1976. The Fine Structure of the Nervous System. W. B. Saunders, Philadelphia.

43. Pickett-Heaps, J. D. 1969. The evolution of the mitotic apparatus: An attempt at comparative ultrastructural cytology in dividing plant cells. *Cytobios. 1*:257-280.

44. Pollard, T. D., and R. R. Weibing. 1974. Actin and myosin and cell movement. *CRC. Crit. Rev. Biochem. 2*:1-65.

45. Porello, K. P., W. Z. Cande, and B. Burnside. 1981. NEM-modified S-1 inhibits reactivated contraction in teleost retinal cone models. *J. Cell Biol. 91*:304 (abs.).

46. Porter, K. R. 1966. Cytoplasmic microtubules and their functions. *In*: Principles of Biomolecular Organization. G. E. W. Wolstenholme and M. O'Connor, editors. J. & A. Churchill, London, pp. 308-345.

47. Porter, K. R., and K. L. Anderson. 1982. The structure of the cytoplasmic matrix preserved by freeze-drying and freeze-substitution. *Eur. J. Cell Biol., 29*:83-96.

48. Porter, K. R., D. P. Boggs, and K. L. Anderson. The distribution of water in the cytoplasm. Paper presented to the 40th Annual EMSA Meeting, Washington, D.C., August 9-13.

49. Porter, K. R., and M. McNiven. 1982. The cytoplast: A unit structure in chromatophores. *Cell, 29*:23-32.

50. Roberts, K. 1974. Cytoplasmic microtubules and their functions. *Prog. Biophys. Mol. Biol. 28*:371-420.

51. Schliwa, M. 1982. Proteins associated with cytoplasmic actin. *Cell 25*:587-590.

52. Schnapp, B. J., and T. S. Reese. 1982. Cytoplasmic structure in rapid frozen axons. *J. Cell Biol., 94*:667-679.

53. Shelanski, M. L., J.-F. Leterrier, and R. K. H. Liem. 1981. Evidence for interactions between neurofilaments and microtubules. *Neurosciences Res. Prog. Bull. 19*:32-42.

54. Shelanski, M. L., and R. K. H. Liem. 1979. Neurofilaments. *J. Neurochem. 33*:5-13.

55. Smith, R. S. 1980. The short term accumulation of axonally transported organelles in the region of localized lesions of single myelinated axons. *J. Neurocytol. 9*:39-65.

56. Snyder, J. A., and J. R. McIntosh. 1976. Biochemistry and physiology of microtubules. *Ann. Rev. Biochem. 45*:699-720.

57. Stearns, M. E., and R. L. Ochs. 1981. A functional *in vitro* model for studies of intracellular motility in permeabilized erythrophores. *J. Cell Biol. 91*:416 (abs.).

58. Steinert, P. M. 1977. The mechanisms of assembly of bovine epidermal keratin filaments *in vitro*. *In*: The Biochemistry of Cutaneous Epidermal Differentiation. M. Seiji and I. A. Bernstein, editors. University of Tokyo Press, Tokyo, pp. 440-451.

59. Tilney, L. G. 1968. Studies on the microtubules in heliozoa: IV. The effect of colchicine on the formation and maintenance of the axopodia and the redevelopment of pattern in *Actinosphaerium nucleofilum* (Barrett). *J. Cell Sci. 3*:549-562.

60. Tilney, L. G., J. Bryan, D. J. Bush, K. Fujiwara, M. S. Mooseker, D. B. Murphy, and D. H. Snyder. 1973. Microtubules. Evidence for 13 protofilaments. *J. Cell Biol. 59*:267-275.

61. Tilney, L. G., and K. R. Porter. 1965. Studies on the microtubules in heliozoa: I. The fine structure of *Actinosphaerium nucleofilum* (Barrett) with particular reference to the axial rod structure. *Protoplasma 60*:317-344.

62. Tsukita, S., and H. Ishikawa. 1976. Three-dimensional distribution of smooth endoplasmic reticulum in myelinated axons. *J. Elec. Micros. 25*:141-149.

63. Tsukita, S., and H. Ishikawa. 1980. Movement of membranous organelles in axons: Electron-microscopic identification of anterogradely and retrogradely transported organelles. *J. Cell Biol. 84*:513-530.

64. Tsukita, S., and H. Ishikawa. 1981. The cytoskeleton in myelinated axons: serial section study. *Biomed. Res. 2*:424-437.

65. Ishikawa, H., and S. Tsukita. 1982. Morphological and functional correlates of axoplasmic transport. *In*: Axoplasmic Transport. D. G. Weiss, editor. Springer-Verlag, Berlin, pp. 251-259.

66. Weisenberg, R. 1972. Microtubule formation *in vitro* in solutions containing low calcium concentrations. *Science 177*:1104-1106.

67. Weiss, P. A., and H. Wang. 1936. Neurofibrils in living ganglion cells of the chick, cultivated *in vitro*. *Anat. Rec. 67*:105-117.

68. Willard, M., W. M. Cowan, and P. R. Vagelos. 1974. The polypeptide composition of intraaxonally transported proteins: Evidence for four transport velocities. *Proc. Natl. Acad. Sci. USA 71*:2183-2187.

69. Wolosewick, J. J. 1980. The application of polyethylene glycol (PEG) to electron microscopy. *J. Cell Biol. 86*:675-681.

70. Wolosewick, J. J., and K. R. Porter. 1976. Stereo high-voltage electron microscopy of whole cells of the human diploid cell line, WI-38. *Am. J. Anat. 147*:303-323.

71. Wolosewick, J. J., and K. R. Porter. 1979. Microtrabecular lattice of the cytoplasmic ground substance: Artifact or reality? *J. Cell Biol. 82*:114-139.

72. Wolosewick, J., and K. R. Porter. 1979. Polyethylene glycol (PEG) and its application in electron microscopy. *J. Cell Biol. 83*:303 (abs.).

Chapter 2

Biochemistry of Neurofilaments

Fung-Chow Chiu, James E. Goldman, and William T. Norton

Neurofilaments are the intermediate filaments of neurons and their processes. They were the first intermediate filament types to be recognized morphologically, and invertebrate neurofilaments were the first of this class of structures to be examined biochemically. In the past decade, work on intermediate filament proteins has progressed rapidly. We now know that most tissues contain filaments of similar morphology but of several chemically distinct types. Neurofilaments remain unusual in this class of structures in that mammalian neurofilaments are composed of three different subunit proteins rather than one. They also appear to be absolutely specific to neuronal cells. On the other hand, filaments of invertebrate axons are different from those in any vertebrate tissue.

It is only recently that techniques have been developed to isolate and purify neurofilaments and to manipulate the subunit proteins. Thus, the biochemistry of filaments is still a young and growing field. In this chapter, we summarize the current knowledge of the isolation, composition, and metabolism of neurofilaments; the evidence for their interactions with microtubules; and their relationship to other intermediate filaments. Since invertebrate neurofilaments are chemically and immunologically distinct, they are discussed in a separate section.

The authors' studies have been supported by U.S. Public Health Service grants NS-02476, NS-03356, NS-17125, and NS-00524. We are grateful for the excellent assistance of Renee Sasso and Marion Levine.

27

Isolation

Many methods have been devised to isolate intact neurofilaments from the mammalian central nervous system (CNS). For convenience, these methods are discussed under four categories: (1) axonal flotation, (2) direct extraction, (3) copurification with microtubules, and (4) cytoskeletal preparation. None of these methods, when applied to the CNS, yields a preparation containing neurofilaments exclusively.

Norton and Turnbull (57) demonstrated that, when CNS white matter was homogenized in isotonic buffer containing 0.32-M sucrose, the myelin sheaths remained attached to the axons. Because of their high lipid content, the myelinated axons formed a floating layer when homogenates were made 0.85-M in sucrose and centrifuged. Thus, more dense contaminants could be removed by repeated flotation. The axons were then stripped free of myelin in hypotonic solutions, and, since the density of neurofilaments is higher than that of myelin, the axonal and myelin components could be separated by centrifugation (22). The first attempts to isolate neurofilaments based on axon-flotation (22,75) yielded preparations that, by electron microscopy, showed a mixture of tightly and loosely packed bundles of 8- to 10-nm filaments. Since astrocyte filaments *in situ* appear closely packed but neurofilaments appear more widely spaced, it was inferred that the tightly packed bundles represented astrocytic filaments and the loosely packed bundles, neurofilaments. Polyacrylamide-gel electrophoresis of such preparations in the presence of sodium dodecyl sulfate (SDS-PAGE) showed a major band of 50,000-mol wt and several prominent bands at higher molecular weight (21). Subsequent modifications of the axon-flotation technique that used a prolonged myelin-stripping step, from 2 hours to overnight, in hypotonic buffer, and repeated centrifugations of the filament pellet (16,71,90) yielded preparations with a more uniform morphology and a simpler biochemical composition than that from the earlier method. Most preparations consisted of loose bundles composed primarily of one major protein of 50,000-mol wt. This led to the incorrect identification of the 50,000-mol wt protein as the subunit of neurofilaments. Later studies (6,49), however, demonstrated that neurofilaments were lost during the prolonged myelin-stripping and centrifugation steps. Furthermore, the tight bundles dispersed during overnight exposure to hypotonic buffer (Chiu, Korey, and Norton, unpublished observation). Thus, the apparent increase in loose bundles did not represent any enrichment in neurofilaments. It is now known that some versions of this procedure furnished an enriched preparation of glial filaments that copurified with the myelinated axons and that the 50,000-mol wt protein was primarily the glial filament subunit, called glial fibrillary acidic protein (GFAP) (27). A detailed history of the mistaken identification of neurofilament subunits has been reviewed elsewhere (56).

Recent studies based on axon flotation have employed a shorter myelin-stripping time. These procedures furnished a mixed population of intermediate filaments containing both neurofilaments and glial filaments (5,7,49). The yields in such preparations have been between 0.2 and 0.5 mg of protein per gram of white matter. The percentage purity of neurofilaments in these preparations, as measured by densitometric scan of stained gels following SDS-PAGE, varied from 23% in calf to 29% in human (7).

The method of direct extraction was first developed by Schlaepfer (68), and the details as applied to peripheral nerves are discussed elsewhere in this volume. (See Chapter 3.) This method originated from the observation that, when nervous tissue was agitated in hypotonic buffer, neurofilaments were released into the suspension. It is not clear whether the neurofilaments were dissolved—all the subunits being in monomeric form—or partially disassembled, since electron microscopy of the suspension revealed filamentous structures. The procedure was later modified to minimize the contamination of soluble protein (69). The suspension was adjusted to 0.1-M sodium chloride and centrifuged at 25,000 rpm to yield a pellet enriched in neurofilaments. When this modified method was applied to rat spinal cord, the yield of total protein in the neurofilament-enriched pellet (SC II) was about 0.6 mg per gram of spinal cord, as calculated from the data of Schlaepfer and Freeman (69). In contrast to the preparation from the PNS, which was contaminated with proteins of 15,000- to 30,000-mol wt, the preparation from spinal cord was contaminated with large amounts of proteins with molecular weights of 45,000 to 55,000.

Variations have been developed to remove the contaminating proteins (54,74). Mori and Kurokawa (54) prepared neurofilaments from sciatic nerves according to the method of Schlaepfer and Freeman (69). The preparation was then centrifuged through a 1.0-M/2.2-M sucrose density gradient. Filamentous materials that collected at the interface between 1.0-M and 2.2-M sucrose were shown to be identical in composition to neurofilaments and mostly free of other protein contaminants. This method has not been applied to the CNS, however. Whether or not the use of a density gradient will be successful in removing the 45,000- to 55,000-mol wt contaminants remains to be seen. A more elaborate procedure was devised by Shecket and Lasek (74) in which fresh spinal cords were homogenized in hypotonic buffer and centrifuged on top of 0.2-M sucrose to remove myelin. The supernates were then treated with 15% ammonium sulfate, centrifuged again to remove precipitated material, and subjected to gel-exclusion chromatography. Neurofilaments, eluted in the void volume as large aggregates, were then precipitated with 40% ethanol. The yield was 0.43 mg protein per gram of spinal cord. The preparations from either the spinal cord or the sciatic nerves were both contaminated with proteins at 46,000- to 57,000-mol wt. Furthermore, this method could be used only with fresh tissue. On the other hand, the advantage of the Shecket and Lasek method is that preparations with comparable yield and purity may be obtained from both the CNS and the PNS.

When microtubules were isolated by an *in vitro* assembly-disassembly method, several other proteins copurified with tubulin. (See, for example, reference 28.) These copurifying proteins were called microtubule-associated proteins (MAPs) (78). Upon the second cycle of assembly-disassembly, some of these MAPs were found to remain in aggregates and could be separated from the disassembled tubulins either by differential centrifugation (20) or by gel-exclusion, chromatography in Bio-gel A-150 (4). Electron microscopy showed that these nondepolymerizable MAPs formed fibers (20). On the basis of their diameter of 10 nm, Berkowitz and others (4) first suggested that these fibers were neurofilaments. In neither of the studies (4,20) was the protein composition of these fibers identified, however. That neurofilament proteins were indeed present in these microtubule fractions became clear when the protein composition of neurofilaments had been established (36,49,68; see also the discussion below). More recently, Delacourte and others have substantially modified their procedure to enhance the yield of neurofilament proteins (19). The purity of neurofilament preparations was greatly improved when the spinal cord, rather than the brain (85), was used as the source of the 10-nm fibers. The yield of this preparation was 1 mg protein per gram spinal cord. This preparation remained contaminated with tubulin and a 50,000-mol wt protein.

The fourth method for neurofilament isolation is an adaptation of a method used to isolate intermediate filaments from non-neural tissues, including avian gizzards and cultured cells (40,79,81). In this technique, tissues or cells were treated with buffers containing nonionic detergents, such as Triton X-100 and NP-40. The cell membranes were solubilized and the cytoplasmic contents released. The materials that remained insoluble, known as cytoskeletons, were composed mostly of intermediate filaments and nuclei.

This method of cytoskeletal preparation was applied to CNS tissue with the addition of a second extraction step in the presence of 0.9-M sucrose (9). The myelin was then effectively removed from the cytoskeletal pellet following centrifugation. Cytoskeletal preparations from spinal cord contained a mixture of neurofilaments and glial filaments as well as some low-molecular-weight proteins, presumably of nuclear origin (Figure 2-1). Gel electrophoresis showed that neurofilament proteins made up 30% of the total protein in this preparation, a degree of purity somewhat less than that found in other neurofilament preparations, such as those derived from direct extraction or microtubule copurification. The major advantage of the cytoskeletal preparation was that the yield, 7 to 10 mg per gram spinal cord, was several times higher than that found using other methods of isolation. Neurofilaments were recovered in apparently quantitative yield: little of the neurofilament proteins were detected in the supernates or myelin fractions (Figure 2-1). Recently, this method of cytoskeletal preparation has been scaled up for the rapid isolation of large quantities of mixed intermediate filaments: 80 to 100 mg of cytoskeletal protein can be prepared from 10 g of rat spinal cords within 1 hour (8). Furthermore, the cytoskeletal preparation may be applied to frozen as well as fresh spinal cord. Unlike the method of axon flotation, which depends on an intact axon-myelin relationship, the integrity of the myelin sheath is not important for cytoskeletal preparation.

Each of the isolation methods mentioned here has its own advantages and disadvantages. The yield and the degree of purity are important considerations when one evaluates these or any other methods of neurofilament isolation. Methods that give low yield would not be useful for small experimental animals or for materials in limited supply. On the other hand, methods that give high yield, for example, the cytoskeletal preparation, are accompanied by a lower degree of purity and, therefore, limit the use of the final preparations if not further purified. Factors that must also be considered include the composition of contaminant materials. For example, the method of axon flotation yields preparations containing large amounts of lipid (57,71), which may interfere with subsequent biochemical studies. Preparations of neurofilaments contaminated with large amounts of glial filaments or tubulin have led to considerable confusion over the interpretation of immunological and biochemical data.

Composition

There is now substantial evidence to support the view that mammalian neurofilaments contain three major proteins with approximate molecular weights of 70,000, 160,000 and 210,000. These three proteins will be referred to as NF proteins and abbreviated 70K, 160K, and 210K. The identification of these three proteins as neurofilament subunits was first proposed by Hoffman and Lasek (36), who demonstrated that these proteins, together with tubulin and actin, were

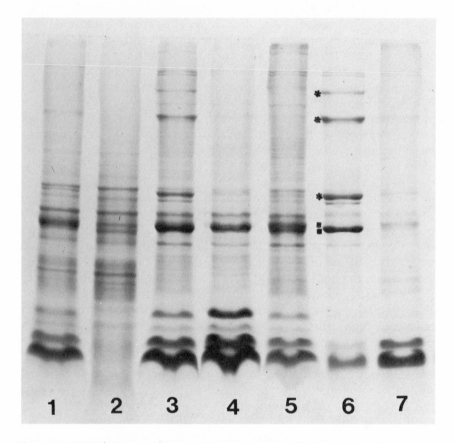

Figure 2-1. SDS-PAGE from CNS cytoskeletal preparation. (1) Homogenates of rat spinal cord, in 0.5% Triton X-100 buffer (9). (2) Supernates after centrifugation at 13,000 g. (3) Pellet after centrifugation at 13,000 g; it contains myelin and cytoskeleton. The pellet (lane 3) was resuspended in 0.9-M sucrose-Triton buffer (9) and centrifuged at 13,000 g to give: (4) a floating layer, containing mostly myelin; (5) a clear supernatant fraction; and (6) a pellet that represents the final product of the cytoskeletal preparation (* marks neurofilament proteins [from top to bottom] 210K, 160K and 70K subunits, ▪ marks glial fibrillary acidic protein). (7) High-speed pellet (100,000-g) of 0.9-M sucrose-Triton supernates (lane 5); note the paucity of neurofilament proteins in the high-speed pellet. (From reference 8. Used with permission.)

transported along axons at a slow rate. Subsequently, DeVries, and others (21) showed that an axon-flotation method yielded preparations that contained glial fibrillary acidic protein as well as three higher-molecular-weight proteins. They postulated that these three proteins were axonal components. Since axon-flotation methods applied to CNS gave fractions containing large amounts of astrocyte intermediate filaments, Schlaepfer and associates (68,69) studied peripheral nerves, which contain no astrocytes. Extraction of nerves yielded a preparation that, by electron microscopy, was shown to contain intermediate filaments with diameters of 10 nm, similar to those of neurofilaments *in situ*. Gel electrophoresis of this preparation revealed large amounts of NF proteins. Other studies have confirmed the existence of the NF proteins in filament preparations from the CNS (4,5,7,14,49).

Additional evidence that supported the identification of the NF proteins as subunits of neurofilaments was found in immunological and pathological studies. Antisera raised against each of the purified proteins react only with axons by immunohistochemical techniques. (See, for example, Figure 2-2 and references 1,49,70,91.) Such antisera have also been found to decorate isolated neurofilaments (70,87).

Changes in neurofilaments induced experimentally have been correlated with changes in the NF proteins. Soifer and others (80) showed that, during Wallerian degeneration, the loss of neurofilaments is accompanied by the disappearance of the NF proteins. Conversely, accumulations of neurofilaments, induced by β, β'-iminodiproprionitrile (34) or aluminum (72) are accompanied by increases in the NF proteins.

A 50,000-mol wt protein found in preparations of CNS intermediate filaments (5,7,16,22,33,90) is now believed to be mostly glial fibrillary acidic protein (27,56). In some preparations, there seems to be a neuronal component in this molecular-weight region since there is some immunological cross-reactivity between material at 50,000- to 57,000-mol wt and antineurofilament antisera. This minor containment is probably derived from degradation of the neurofilament proteins (1,77).

To summarize, the NF proteins were first found in preparations enriched in neurofilaments. Immunocytochemical studies localized these proteins to axons and isolated neurofilaments. Therefore, by inference, the NF proteins are components of neurofilaments. The neuropathological data confirmed this interpretation. Whether or not neurofilaments are composed exclusively of the three NF proteins remains to be proven.

The molecular weights of neurofilament proteins vary slightly from one species to another. Although some of the reported variation may be attributed to the differences in the analytical techniques used in different laboratories, the phenomena of interspecies variation in molecular weight is nevertheless real (5,7,49). Gel electrophoresis of intermediate filaments from three different species—human, rat, and bovine (Figure 2-3)—when analyzed side by side, clearly show that within one molecular-weight group, such as the 70K subunits, each

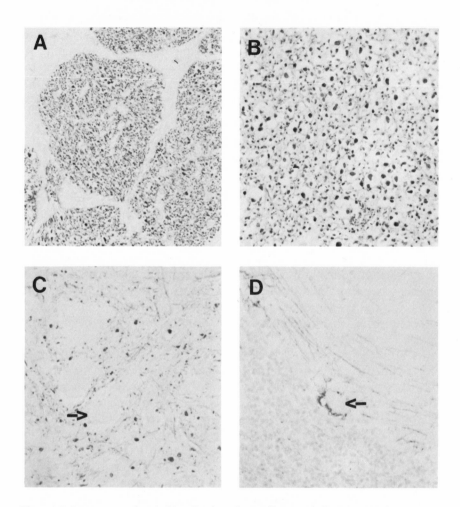

Figure 2-2. Immunocytochemical localization of neurofilaments in human CNS through the use of a mouse monoclonal antibody raised against the 210K neurofilament protein isolated by column chromatography. (See Figure 2-4.) This antiserum cross-reacts with the 150K neurofilament component. Binding of the primary antiserum to tissues obtained at autopsy was visualized with an avidin-biotin technique on paraffin sections. Positive reactions are seen in (A) optic nerve fibers, (B) axons of dorsal columns of the spinal cord, (C) axons in the anterior horns of the cord, and (D) Purkinje cell baskets in the cerebellum. Neither anterior horn cells nor Purkinje cell perikarya show reaction product (arrows). (Sections were counterstained with methyl green, and all photographs are magnified ✕250.) (Courtesy of Drs. Allen Gown and Arthur Vogel, University of Washington School of Medicine.)

protein may have slightly different migration rates. However, peptide analyses have demonstrated that each protein, despite its variation in molecular weight, is nearly identical from one species to another. For example, the 70K protein from rat, mouse, human, and bovine all show similar peptide maps (5,7). Furthermore, within one animal, the molecular weights of the NF proteins are consistent throughout different parts of the nervous system (11,14).

On the other hand, biochemical relationships among the NF proteins are not well understood. Peptide analysis with two different mapping techniques indicates that no oligomeric relationship exists among the proteins (7). In particular, it is unlikely that the 210K protein is a trimer of the 70K protein or that the 210K protein is cleaved to yield the 160K and 70K subunits. Czosnek and others (13) demonstrated that all three NF proteins were synthesized *in vitro* from spinal-cord mRNA. They suggested that these proteins are derived from different genes and are, therefore, synthesized independently. Preliminary studies (Chapter 4) showing differential expression of the NF proteins during development tend to support this view.

Recent data of amino acid analysis of the NF proteins also do not indicate any oligomeric or precursor relationship (8,37). These data will be discussed below. It is sufficient to point out that, although the overall contents of acidic and basic residues in all of the NF proteins are similar (21), there are significant differences in the amino acid compositions (Table 2-1). For example, if the 210K protein were degraded to yield the 70K and 60K proteins, its proline content would be a weighted average of the proline contents of the 70K and 160K subunits. This is not the case.

Immunological studies have demonstrated that the NF proteins share common antigenic determinants. Antisera raised against the individual proteins cross-react with the other proteins (1,49,70,87,90). By selective adsorption with purified proteins, antisera specific to only one protein have been obtained (70,87). The biochemical and immunological data together suggest that the NF proteins, although sharing some common amino acid sequences, are different from each other. The precise structural relationships among the neurofilament subunits can be elucidated only by future studies of the primary sequences of these proteins.

The stoichiometry of the NF proteins has been described in several studies (5,7,12,54,74). It has been determined by densitometric scans of NF proteins separated by SDS-PAGE and varies from one laboratory to another, ranging from 9:2:1 (54) to 6:2:1 (74) for the ratio of 70K:160K:210K. One reason for this variation is that different investigators used different dyes for gel staining. The individual proteins do not stain uniformly; in particular, the 210K is underrepresented by as much as 50% with Coomassie-blue staining (8). When stained with fast green, neurofilaments isolated by two very

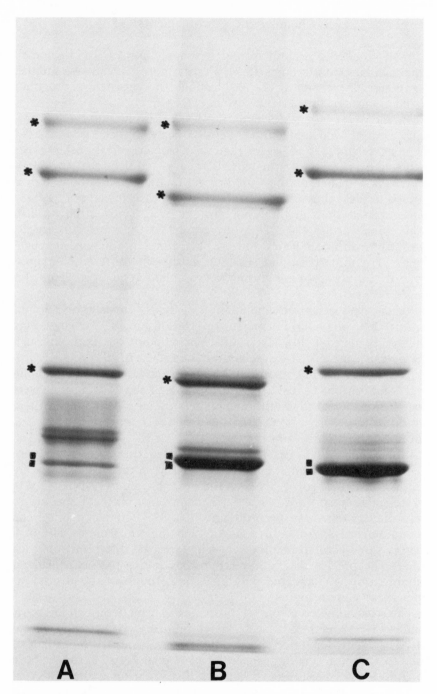

Figure 2-3. SDS-PAGE of CNS intermediate filaments from (A) human, (B) rat, and (C) bovine. (* Neurofilament proteins [from top to bottom] 210K, 160K and 70K; ∎ glial fibrillary acidic protein.) Filaments were prepared with the method of axonal flotation (7).

Table 2-1.
Amino Acid Analysis of Neurofilament Proteins
(Mole %): Mean ± Standard Deviation

	70,000	160,000	210,000
Asp	7.98 ± 1.12	6.68 ± 0.17	5.07 ± 0.48
Thr	3.62 ± 0.24	4.10 ± 0.36	4.63 ± 0.85
Ser	9.87 ± 0.82	8.93 ± 0.52	8.54 ± 0.46
Glu	22.46 ± 0.39	24.57 ± 0.38	20.00 ± 0.65
Pro	3.02 ± 0.11	4.55 ± 0.21	8.53 ± 0.37
Gly	5.31 ± 1.18	5.63 ± 0.25	5.23 ± 0.39
Ala	10.66 ± 0.70	8.77 ± 0.11	14.48 ± 0.89
Val	4.46 ± 0.09	6.28 ± 0.11	5.75 ± 0.14
Met	1.85 ± 0.02	0.96 ± 0.40	0.95 ± 0.39
ILe	3.23 ± 0.08	3.33 ± 0.09	2.04 ± 0.22
Leu	9.10 ± 0.50	6.81 ± 0.31	6.30 ± 0.41
Tyr	3.17 ± 0.22	1.75 ± 0.01	1.16 ± 0.22
Phe	2.20 ± 0.03	1.59 ± 0.08	1.50 ± 0.29
His	1.11 ± 0.06	1.74 ± 0.03	1.36 ± 0.24
Lys	6.73 ± 0.25	10.25 ± 0.43	11.00 ± 0.53
Arg	6.47 ± 0.45	4.79 ± 0.27	4.79 ± 0.18
Cys	0.40	0.49	0.77
Trp	0.30	0.13	0.11

Source: Reference 8. Used with permission.
Note: Amino acid compositions were measured in 70,000 and 160,000 mol wt from three different preparations; in 210,000 mol wt, from four different preparations. Measurements of cysteine (as cysteic acid) and tryptophan were made from one preparation. No corrections have been applied for any hydrolytic destruction of serine and threonine.

different methods—direct extraction and cytoskeletal preparation—show a similar stoichiometry of approximately 6:2:1 (8,74). This indicates that the relative amounts of NF proteins remain unchanged during isolation. Variable stoichiometries may also be due to real interspecies differences. For example, neurofilaments from calf appeared to contain less 210K than neurofilaments from rat (7). The significance of the stoichiometry among the NF proteins will only be fully appreciated after the structure of neurofilaments has been described in far more detail.

Separation of Subunits and Amino Acid Composition

The subunits must be separated before the biochemistry of the individual NF proteins can be studied. Until recently, this has been accomplished only with preparative SDS-PAGE (7,21,37). This method of separation has two drawbacks: low yield and the presence of detergent in the purified proteins. Through the use of the cytoskeletal preparation, the problem of low yield has been circumvented. The presence of SDS, however, has remained an obstacle to further studies, such as the measurement of interactions among the subunits. Clearly, a method to separate the individual proteins without the use of SDS is desirable.

Two new approaches have been taken for the separation of neurofilament subunits in the absence of detergent. One approach, which uses ion-exchange chromatography, was developed independently by two groups (29,48). The other approach took advantage of the differences in molecular weight among the NF proteins and separated the subunits by gel-filtration chromatography (8). Both approaches share two common aspects: (1) neurofilaments were first isolated and used as crude materials for chromatography (it is not possible, at present, to isolate the individual proteins directly from tissue homogenates); (2) neurofilaments, once isolated, were solubilized in buffers containing 8M urea prior to chromatography. The degree of resolution and the percentage of protein recovered after chromatography are two important considerations when these or any other methods of separation are evaluated. Starting with 50 g of white matter, Liem and Hutchison obtained 10 mg of neurofilament proteins, which they subsequently separated into 1 mg of the 210K protein, 0.5 mg of the 160K protein, and 2 mg of the 70K protein (48). The combined yield of purified subunits was, therefore, 0.07 mg per gram. The gel-filtration method, as used in our laboratory, has allowed excellent recovery of proteins. When the method was properly executed, no protein was retained in the column. However, there were fractions that contained overlapping boundaries of the 160K and 210K proteins. The typical yield of the individual neurofilament proteins, as separated by gel filtration of cytoskeletons, was about 0.4 to 0.6 mg per gram of rat spinal cord. The combined yield of the purified subunits was, therefore, about 1.5 mg per gram of tissue. This represented 50% to 60% recovery of the total neurofilament proteins applied to the gel-filtration column. The recovery may be improved with rechromatography of the overlapping fractions. Individual neurofilament proteins may be freed of urea and salt by dialysis and remain stable in solution (Figure 2-4). The degrees of purity of the proteins were acceptable for further biochemical studies, such as amino acid analysis.

The amino acid compositions of the neurofilament proteins were first described by DeVries and others (21). Intermediate filaments from bovine CNS were resolved on SDS-PAGE, and the three neurofilament proteins were eluted from the gel. The molecular weights of

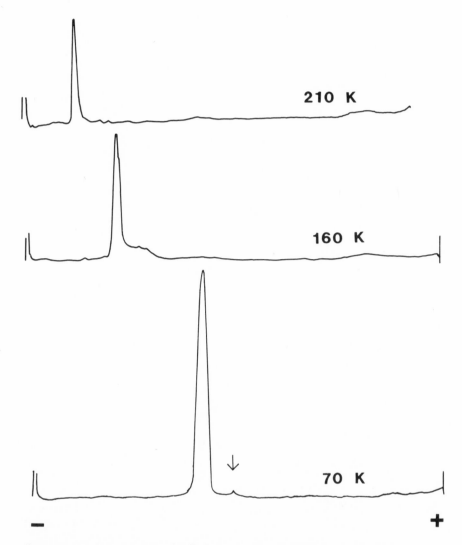

Figure 2-4. Densitometric scans of individual neurofilament subunits after SDS-PAGE. Neurofilament proteins were separated by gel filtration, dialyzed free of urea and salt overnight, electrophoresed on a 5% to15% polyacrylamide gradient gel, and stained with Coomassie blue. Arrow on the 70K scan indicates the position of glial fibrillary acidic protein. (From reference 8. Used with permission.)

the 160K and 210K proteins were underestimated, since the electrophoretic system used in this study was best suited for resolution at less than M_r-100,000. Nevertheless, the neurofilament proteins were found to contain about 30% acidic residues (aspartic and glutamic acids) and 15% basic residues (lysine, arginine, and histidine). DeVries and others (21) concluded that all of the neurofilament subunits are acidic proteins. Recent studies on neurofilaments from rat CNS (8) and bovine CNS (37) generally confirmed this observation. In addition, more detailed analysis was possible because of improvement in the separation techniques. Data presented in Table 2-1 indicate that, although the 70K and 160K proteins both contained 30% acidic residues, the 210K protein contained only 25%. This is consistent with the reported isoelectric points (pI) of about pH 5 for the 70K and the 160K proteins and a somewhat more basic pI for the 210K protein (5,10,85).

The proline content in the 210K protein, at 8.5%, is much higher than that in the 70K and 160K proteins. This is an interesting finding since a molar percentage of 8.5 represents more than 15 residues of proline in a protein with a molecular weight of 210,000. It raises the possibility that the secondary structure of the 210K protein may be quite different from that of the 70K and 160K proteins. There are other important differences in the amino acid compositions: methionine and tyrosine in the 70K protein are both higher than those in the other two subunits; cysteine (as cysteic acid) in the 210K protein is higher than that in the 70K and 160K proteins. Although there is an apparent difference in the molar percentage of tryptophan among the NF proteins, the number of tryptophan residues, at two per protein, is the same for all three subunits.

Solubilization and Reassembly

Davison and Winslow (16) first reported that CNS intermediate filaments can be solubilized in guanidine hydrochloride. More recently, urea has been used to dissolve neurofilament preparations for the purposes of two-dimensional gel electrophoresis (5,10,85) and chromatography (8,29,47,53). That the neurofilaments were completely dissolved was evident because the subunit proteins eluted from gel-filtration columns at rates appropriate to their individual molecular weights (8).

Davison and Winslow (16) observed the formation of 10-nm filaments when the chaotropic agent was removed. The major protein in that preparation was 53,000- to 55,000-mol wt, however, and was probably glial fibrillary acidic protein (GFAP). Liem (47) recently purified neurofilament preparations by hydroxylapatite chromatography in urea. The major contaminant, GFAP, was not retained by the column; the neurofilament proteins were retained and could be eluted with 0.3-M sodium phosphate. Removal of urea yielded filaments 10 nm in diameter, which contained the NF proteins.

Since urea-solubilized neurofilaments can be separated into individual subunits (see above), their individual assembly properties can be examined. Thus, dialysis of the 70K subunit against a buffer containing MES (2-[N-morpholino] ethanesulfonic acid) and 0.17-M sodium chloride resulted in assembly of smooth 10-nm filaments (29). On the other hand, neither the 160K nor the 210K subunit was capable of self-assembly under this condition (29). These pre-liminary findings seem to be consistent with the postulate of Willard and Simon (87) that the 70K subunit occupies the central core of the neurofilament whereas the 210K subunit is located in a peripheral position. At present, the interpretation of the reassembly data is somewhat tentative. The effect of urea denaturation on the subsequent reassembly of neurofilament subunits is unknown. Re-assembly under other conditions—with evaluations of the require-ments for ions, ionic strength, and energy sources—needs to be ex-plored. The assessment of reassembly of mammalian neurofilaments has been restricted to electron microscopy, which yields mostly qual-itative information. In order to evaluate fully various conditions of reassembly, biophysical and biochemical methods will be necessary to obtain quantitative data, not only on reassembly but also on pos-sible interactions among the neurofilament proteins that electron microscopy does not reveal.

Metabolism

Synthesis and Turnover

The aspects of synthesis and turnover in the biochemistry of neuro-filaments have received little attention. Somewhat more effort has been devoted to studies of their degradation, which is discussed in Chapter 5 of this volume.

The immunological cross-reactivity among the three neurofilament proteins indicates that they share common sequences (see above). This suggests either that the 210K and 160K species were oligomers of the 70K subunit or that the two smaller proteins were derived from the larger one by posttranslational proteolysis. A third possibil-ity that all three proteins were generated by cleavage of a single large precursor of 410,000- to 450,000-mol wt was also suggested (77). All three of these possibilities required that the three subunits were products of a single gene.

The various mapping and compositional studies, summarized above, indicated that the three proteins were neither related by oli-gomerization nor derived from the 210K protein. Czosnek and others (13) have now shown that all three neurofilament proteins were among the proteins synthesized when polyribosomes from rabbit spinal cord were translated in a rabbit reticulocyte lysate system.

These data indicate that each neurofilament protein is the product of a separate gene. This same study also showed that the translation products of a homologous cell-free system of rabbit spinal cord were very similar to those synthesized by rabbit cord *in vivo* and that all three neurofilament proteins were made in each system. Apparently, under most circumstances, the synthesis of all three proteins is co-ordinated. Since each is a separate gene product, however, it is possible that any one protein or combination of proteins could be generated independently. There is evidence that the three proteins appear at different times in the developing nervous system. (See Chapter 4.)

Neuronal perikarya, isolated from young rat brain, have also been used for biosynthetic studies (55). If prepared properly, with a minimum of trypsin, these cells are capable of protein synthesis for up to 90 minutes. The newly synthesized proteins are primarily actin and tubulin, but smaller amounts of the neurofilament triplet are also made. The identification of several *in vitro* systems capable of neurofilament synthesis should lead to more extensive studies of their metabolism.

All three neurofilament proteins are readily labeled when radioactive amino acids are injected into the spinal cord, dorsal root ganglia, or retina. This approach, which forms the basis of axonal transport studies of filaments and other structures (see Chapter 4), indicates that neurofilaments are continually synthesized in the adult animal. The bulk of the filaments are then transported down the axon in the slow component (SCa) of axoplasmic flow at rates of 0.3 to 1.0 mm per day. Since this is a continuous process, even in axons that are not growing, the neurofilament proteins must eventually be degraded at a rate equal to their rate of export into the axon.

Weiss and Hiscoe (86) suggested that transported proteins must be degraded throughout the length of the axon; otherwise, they would pile up at the terminal. If there were no protein synthesis in the axon, this mechanism would require that there be a continually diminishing amount of transported elements as they progressed distally. The density of neurofilaments and the average diameter of an axon, however, are fairly constant throughout much of the length of an axon (reviewed in 56). Therefore, since apart from mitochondrial activity there appears to be little local protein synthesis in the axon, Lasek and Hoffman (43) proposed that axonal neurofilaments do not undergo degradation until they reach the presynaptic terminal where catabolism could take place.

Support for this mechanism comes from quantitative studies, in the guinea pig phrenic nerve, of the transported wave of radioactivity in SCa, which contains neurofilaments and microtubules. Lasek and Black found that, after injection of labeled precursor into the ventral horn, there was the same amount of radioactivity in this wave after 64 days as after 32 days (42). Thus, there was no loss of labeled protein for 32 days, during which the proteins moved 30 mm, confirming previous qualitative observations that radioactivity in SCa showed no significant attenuation over long periods. Similar studies of transport in the guinea pig optic nerve showed that, once the components of SCa entered the presynaptic terminals (superior colliculus), they were rapidly degraded (42). (See also Chapter 5).

These studies indicate that the turnover times of axonal neurofilaments are a function of their transport rate and axonal length. Neurofilaments in an axon 1 mm long would have a lifetime of a day or two, but those in an axon 1 m in length would be stable for 3 to 10 years. These arguments apply equally well to microtubules in the axon or, in fact, to any axonal proteins undergoing anterograde transport that are not deposited en route to the terminal. There are as yet no studies of turnover of neurofilament proteins in brain. One is not justified, however, in extrapolating these findings to whole brain and deducing average half-lives of cytoskeletal proteins from average axonal lengths. There may well be pools of neurofilaments that are not transported and turn over rapidly. Furthermore, large nonaxonal pools of microtubules exist both in neurons and other cell types.

Phosphate Metabolism

Phosphorylated species of intermediate filament proteins exist in several cell types. (See reference 45 for a review.) Neurofilaments are also phosphorylated, and one of the kinases that can carry out this reaction *in vitro* is associated with microtubules. There is also evidence that neurofilaments themselves, or enzymes closely associated with them, can mediate certain reactions of phosphate metabolism. These aspects of neurofilament metabolism, in conjunction with the morphological and biochemical evidence that neurofilaments and microtubules are closely associated, have received much attention recently. The presumption is that these structures are not static cellular scaffolds (as the term "cytoskeleton" implies) but that through their interactions they play a dynamic role in neuronal metabolism.

Partially purified neurofilaments from guinea pig peripheral nerve (73) and from bovine brain (65) contain an endogenous cAMP-independent kinase capable of phosphorylating the three neurofilament proteins. The kinase in the brain preparation, which appeared to be firmly associated with the filaments, could also phosphorylate the high-molecular-weight microtubule-associated protein, MAP-2, and the addition of MAP-2 stimulated the phosphorylation of the neurofilament proteins. Crude microtubule preparations that contained neurofilaments have two kinase activities, one cAMP independent and one cAMP dependent. When the neurofilaments were separated from microtubules by gel-filtration chromatography, the cAMP-dependent activity stayed with the microtubules, whereas the cAMP-independent activity appeared to remain associated with the neurofilaments (65).

Since the neurofilaments are phosphorylated *in vivo* (40), it is possible that this reaction could be carried out by the kinase that co-purified with the isolated filaments. When rat brain neurofilaments were phosphorylated *in vivo*, the 210K subunit was labeled much more than either the 160K or 70K subunits, whereas the 160K and 70K proteins were more heavily labeled than the 210K protein by the endogenous kinase *in vitro* (40). Peptide mapping studies showed that the 70K protein was phosphorylated at the same sites both *in vivo* and *in vitro*. The 160K protein, however, was labeled at extra sites *in vivo*, whereas the 210K protein was phosphorylated at extra sites *in vitro*. These data indicate that the kinase that copurifies with the neurofilaments could be involved with their phosphorylation *in vivo* but do not prove that postulate unambiguously.

The significance of the neurofilament-associated kinase is diminished by the recent finding that purified neurofilaments, from either bovine brain or rabbit spinal roots, lacked protein-kinase activity for neurofilaments or for an exogenous protein acceptor (46). Neurofilament proteins, however, could be phosphorylated if microtubule proteins were added. Under these conditions, the tau proteins, high-molecular-weight MAPs, and tubulin in the microtubule preparation were also phosphorylated. This kinase activity was cAMP dependent and predominantly phosphorylated the 160K species of the neurofilament triplet.

Two other enzymes of phosphate metabolism—a $3'$, $5'$-cyclic nucleotide, $5'$-phosphodiesterase (63) and a Ca^{++}, Mg^{++}-stimulated ATPase (76)—have been reported to be present in mammalian neurofilament preparations. The cyclic nucleotide phosphodiesterase activity was found in a brain microtubule preparation that had previously been shown to contain 10-nm filaments as well as microtubules (4). When the filaments were separated from the other proteins by gel-exclusion chromatography, the enzyme activity was mostly recovered in the 10-nm filaments (63). A preparation of total brain filaments, isolated directly by the method of Runge and others (62), also contained the phosphodiesterase but at a lower specific activity. The enzyme could not be dissociated from the filament fraction by extraction with 0.5-M salt. This phosphodiesterase activity is activated by calmodulin, and a calmodulin-like activator could be released from brain intermediate filaments on boiling.

Recently, Glicksman and Willard (32) confirmed the results of

Shelanski and others (76) that neurofilament preparations contained ATPase activity. This was interesting because they used the isolation method of Schlaepfer and Freeman (69), which involves partial solubilization of the filaments in hypotonic buffer, whereas Shelanski and others (76) presumably used filaments isolated by a variation of the axon-flotation method. This ATPase activity, however, was not inhibited by antisera to neurofilament proteins and remained soluble when 90% of the neurofilament protein was immunoprecipitated by antibody (32).

Of the three enzyme activities — protein kinase, ATPase, and cAMP-phosphodiesterase — reported in neurofilament preparations, the protein kinase and ATPase have been shown to be readily dissociated, under the proper conditions, from neurofilaments themselves and, therefore, not to reside in any of the neurofilament subunits. These results show that considerable caution must be exercised in interpreting the significance of neurofilament-associated enzyme activities. All of the methods currently available for isolating neurofilaments yield preparations that contain significant amounts of a variety of other proteins whose composition will depend on the preparative method. Cytoskeleton-associated enzymes may eventually be discovered that have a role in neurofilament function, but distinguishing those truly associated with neurofilaments from those adventitiously present in the preparations promises to be a difficult task.

Interactions with Microtubules

Morphological studies show that neurofilaments are linked to each other by sidearms that maintain a loose array with approximately 30 to 40 nm spacing between filaments. Similar sidearms also appear to connect neurofilaments to microtubules (56,89). Recent ultrastructural investigations of cytoskeletons in the axon and in other cell types show the existence of a continuous network of interconnected microtubules, intermediate filaments, and microfilaments. (See Chapter 1.) The well-known phenomenon of the "collapse" of the intermediate filament network in cultured cells on addition of mitotic spindle inhibitors such as colchicine also suggests that organization of many types of intermediate filaments is maintained by an interaction with microtubules. Similar phenomena are seen in nerve cells, where neurofilament bundles become very prominent in the perikarya in animals treated with spindle inhibitors (88). In addition, the fact that only neurofilaments and microtubules move together in a slow component of axonal transport (36; Chapter 4) is evidence for a special relationship between these two structures in the axon.

Biochemical evidence for the interaction of neurofilaments and microtubules has been difficult to obtain. Several investigators have noted that brain microtubules, purified through several cycles of assembly-disassembly, contain 10-nm filaments and neurofilament triplet proteins (4,62,85). Microtubules can be prepared from brain that do not contain neurofilament proteins (77), and neurofilament preparations usually do not contain tubulin, unless isolated from frozen brain (7). The 10-nm filaments that copurify with some microtubule preparations can be separated from microtubules, and it has been noted that these neurofilaments are not contaminated with glial filament protein, whereas brain filaments isolated by axonal flotation do not contain tubulin but do contain large quantities of glial intermediate filaments (77,85).

These results seem rather confusing but can now be better understood with our increased knowledge of the properties of neurofilaments (see above). Neurofilaments tend to disassemble in low ionic strength buffers, whereas glial filaments do not. The disassembled neurofilaments are probably not monomeric but exist as small protofilaments. Brain filaments isolated by axon flotation have never disassembled, and the soluble tubulin and disassembled filaments are discarded in the various supernates. On the other hand, during microtubule isolation, the first extract of disassembled soluble microtubule proteins will also contain some nonsedimentable protofilaments. During the assembly of microtubules, the protofilaments also assemble into larger filaments and the two structures coprecipitate. During the next cycle of cold disassembly of microtubules, the filaments tend to remain insoluble and have been called by some the "cold non-depolymerizable fraction" (19). In this preparation, the insoluble brain filaments, which include essentially all of the glial filaments, would have been discarded in the first extraction step. Therefore, the presence of 10-nm neurofilaments in partially purified microtubules is not strong evidence for a specific interaction between them. More probably, these two structures copurify through a few cycles of assembly-disassembly.

Much stronger evidence for interaction has come from studies of the addition of neurofilaments to microtubules and microtubule proteins (64,76). Runge and others (64) showed that neurofilaments and microtubules (or purified tubulin) would form a complex, as measured by an increase in viscosity, when ATP was added. They speculated that the protein kinase in the neurofilaments (see above) phosphorylated one of the proteins in the system, which initiated complex formation. Shelanski and associates (76) also found that isolated neurofilaments form a complex with microtubules. Their studies indicate that both high-molecular-weight MAPs and tau protein will bind to neurofilaments but that purified 6S tubulin does not bind.

Moreover, if microtubule proteins are polymerized in the presence of neurofilaments and then depolymerized and the filaments removed, the high-molecular-weight MAPS are diminished and the assembly of microtubules is impaired. These data indicate that neurofilaments do react with the components believed to comprise the sidearms of the microtubules.

These two studies (64,76) indicate that neurofilaments and microtubules can interact *in vitro*. That complex formation is mediated by ATP (64) and that one component of neurofilaments is phosphorylated by a cAMP-dependent microtubule-associated kinase (46) strongly suggest that the interaction is regulated by protein phosphorylation of some sort. This view is supported by the possible association of a protein kinase and a cAMP-phosphodiesterase with neurofilaments and by the fact that all three neurofilament proteins are phosphorylated *in vivo*. The nature of this interaction and its metabolic role in the neuron are as yet only matters for speculation.

Invertebrate Neurofilaments

Because of their large size, axons of several invertebrates have been popular experimental preparations. Besides being used in pioneering studies of axonal conduction, invertebrate axons were also subjects of early studies of the chemical composition and physical properties of axoplasm (2,15,18,50). These studies were performed before there was detailed knowledge of the protein constituents of axoplasm, but they suggested the presence of a highly ordered, gel-like axoplasm, with prominent filamentous structures oriented longitudinally. Early work on the effects of ionic strength and pH in altering axoplasmic properties are now being reevaluated in light of what is known about axoplasmic proteins, especially neurofilaments.

Two organisms of particular attractiveness have been the squid (*Dosidicus gigas*, or *Loligo pealii*) and the marine polychaete (*Myxicola infundibulum*). Both contain giant axons, from which axoplasm can be readily extruded in a short time (seconds, in fact, from *Myxicola*). From early studies the gel-like properties of axoplasm were known. Electron microscopy of isolated axoplasm revealed long filamentous structures, 8 nm in diameter (15). Isolated axoplasm also displayed a high-flow birefringence, a property that was lost when axoplasm was dispersed by changes in pH or ionic strength (15). Both pH and ionic strength appear to be factors. For example, squid axoplasm retained high-flow birefringence at low ionic strength (sodium chloride less than 0.3 M) up to pH 8 but dispersed at higher ionic strength at neutral pH (15,38). The stability of *Myxicola* axoplasm is similarly affected by high salt concentration (30). Chaotropic agents, such as urea and guanidine hydrochloride, disperse axoplasm effectively.

Exactly what events take place during these alterations in axoplasmic properties is not yet known, but it is clear that important changes in neurofilament organization take place. For example, axoplasm dissociated by urea shows a preponderance of slowly sedimenting units during analytical ultracentrifugation, but it shifts to larger units upon dialysis against 0.1-M sodium chloride (38). *Myxicola* neurofilaments also shift from fast- to slow-sedimenting material as sodium chloride concentration is increased from 0.2 to 1.0 M (3). Sedimentation peaks appeared hypersharp, suggesting asymmetric units. Krishnan and others (41), in examining *Myxicola* filaments by electron microscopy during treatment with high potassium chloride concentrations or urea, described several stages of dispersal. During early stages, two filaments 4 to 5.5 nm in diameter and wound about each other could be resolved. Occasionally, thinner filaments, 2.0 to 2.5 nm in diameter, were observed. In later stages, long polymers apparently had been broken up into shorter segments, with either a rod-shaped or a globular morphology. Dialysis against a solution of 10 mM Tris-HCl buffer resulted in formation of filamentous structures. Squid brain filaments, dispersed by 1-M potassium chloride, appear as short structures 3 to 5 nm in diameter (92). Thus, high salt and chaotropic substances cause disruption of filaments into small fragments. There is no evidence from these studies that filaments are dispersed into monomeric units. (See the earlier discussion in this chapter for evidence that urea will dissociate mammalian neurofilaments into individual subunits.)

Several orders of helical organization have been inferred in invertebrate neurofilaments. The subunits themselves contain high levels of α-helical content, assessed by circular dichroism and X-ray diffraction studies in *Myxicola* (17), similar to significant α-helical content of other intermediate filaments (82). Individual units (of asymmetric configuration, at least in *Myxicola*) are organized into polymeric strands, their long axes oriented longitudinally, since a molecule 2.4 by 250.0 nm must be oriented this way in a structure of 10-nm diameter. Two or more of these protofilaments are wound about each other. The helical pattern and size are supported by Metuzals's (51) measurements, by Gilbert (30), and by Krishnan and others (41). The fully formed neurofilaments are oriented obliquely to the long axis of the axon, forming a helix of large scale (30,52). An intrinsic twist to the filament protomer may well be responsible for the more complex, larger helical organizations.

The first attempt at isolation and chemical characterization of neurofilament proteins was reported by Huneeus and Davison (38).

Dispersed axoplasm was treated with urea or guanidine hydrochloride, both of which caused a disappearance of filamentous structures, as judged ultrastructurally. Dialysis against a low-ionic-strength salt solution caused filamentous structures to form. A major protein of the re-formed filaments was characterized as a component of about 74,000 mol wt, which contained high proportions of aspartic and glutamic acid residues. A shift from small to large size by dialyzing out urea or guanidine was detected by ultracentrifugation. Sedimenting components in guanidine showed a high frictional coefficient, which Huneeus and Davison ascribed to hydration or asymmetry.

More recent work on squid filaments (44) has produced a cleaner preparation of neurofilaments. When axoplasm was dispersed in 0.1-M sodium chloride or potassium chloride filaments settled in a 1-M sucrose cushion at a 1/2.5-M sucrose interface during density-gradient centrifugation. The filament fraction was composed largely of two proteins of 200,000 (200K) and 60,000 (60K) mol wt. The smaller one is probably the filament protein characterized by Huneeus and Davison (38).

Myxicola neurofilaments have molecular weights different from both squid and mammalian proteins. Major axoplasmic proteins have molecular weights of about 160,000 and 150,000 (31). These molecular weights have been confirmed by filament isolation on density gradients (44) and by precipitating filaments with cytochrome c (25, 31), in the presence of which ring structures are formed. The most recent report gives the molecular weights as 172,000 (172K) and 155,000 (155K) based on electrophoretic migration (25). In addition to the two major high-molecular-weight components, neurofilament preparations from *Myxicola* also show a number of smaller proteins. Recent peptide-mapping experiments with trypsin and papain have demonstrated that the two major proteins are closely related to each other and to a number of smaller proteins (25). For several reasons, including the rapidity of dissection, it is likely that these smaller species represent proteolysis *in vivo* rather than degradation during extraction procedures and that the major 155K protein is derived from the 172K component. These extensive mapping studies provide a solid foundation for a detailed examination of neurofilament structure.

Neurofilament purification from squid brains by disassembly in high-salt (1-M potassium chloride) and reassembly at low-ionic strength has been attempted (92). A major protein of 60K and several minor components, including one of 220K, were isolated. It is possible that proteins other than the 60K and 220K originated from nonneuronal cells. Growth of long filaments from the short units, derived from high-salt dispersal, took place quickly and was conveniently followed by turbidimetric measurements.

The rapid dispersal (loss of gel-like properties) of squid and *Myxicola* axoplasm in the presence of calcium (30,35) is due, at least in part, to neurofilament degradation. Gilbert (30) originally demonstrated a rapid loss of *Myxicola* neurofilament proteins in axoplasm exposed to calcium. Loss of neurofilaments, which was accompanied by the appearance of lower-molecular-weight components on polyacrylamide gels, was not observed in the absence of calcium and could be prevented by several protease inhibitors. Further work extended these observations for squid axoplasm (58,59). The enzyme activity was clearly stimulated by calcium and inhibited by several compounds that block thiol protease activity. Other divalent cations, including Mg^{++}, Co^{++}, and Ba^{++}, did not stimulate proteolysis. Although a number of proteins in squid axoplasm are degraded in the presence of calcium, the 210K neurofilament component appears to be one of the most susceptible. It is not known how many enzymes are responsible for filament proteolysis. (See Chapter 5.)

The Ca^{++}-stimulated proteolysis in *Myxicola* axoplasm produces several neurofilament-degradation products that can be seen in freshly isolated axoplasm (24). This suggests that the proteolytic activity may function *in vivo* within the axon. Other neurofilament peptides smaller than the parent molecules do not appear to be generated by a Ca^{++}-stimulated activity, however. Calcium-activated proteolysis in *Myxicola* produces polypeptides that remain in supernates during sedimentation at high-g forces and polypeptides that are sedimentable (24). It is interesting that the major sedimentable products are smaller than the soluble ones; this raises the possibility that portions of the molecule involved in intermolecular associations (i.e., filament formation) may be localized to a relatively small segment of the protein.

Like mammalian neurofilaments, invertebrate neurofilaments can be phosphorylated. Incorporation of phosphate from $\gamma-^{32}P$ ATP into neurofilament proteins occurs in squid (60) and *Myxicola* (24,44). Peptide-mapping experiments should lead to a detailed picture of where phosphorylated residues are located (24).

One immunocytochemical study of invertebrate neurofilaments has been published; it uses a monospecific antiserum raised in rabbits against *Myxicola* neurofilaments (26). This antiserum reacted with cell bodies and axons but not with glia in tissue sections at both light- and electron-microscopic levels.

Comparison of Neurofilaments with Other Intermediate Filaments

Five classes of intermediate filaments have been described, the subunits of which can be distinguished immunologically and biochemically.

(See reference 45 for a review.) Each class has a different pattern of tissue distribution. Thus, GFAP has been found in CNS glia, both astrocytes and ependyma, keratin in epithelial cells, and desmin in muscle. Vimentin has the widest distribution, having been found in cells of mesenchymal derivation, in a variety of cells in tissue culture (including epithelial cells and glia), and in a number of cell types during development. Neurofilaments have so far been localized exclusively to neurons, both CNS and PNS, and can be detected immunologically early in the development of the nervous system (83,84).

Neurofilaments share a number of biochemical features with other intermediate filament proteins but are clearly distinguishable from them in several ways. Neurofilaments display sidearms, linking them to each other and to other axonal organelles, while other intermediate filaments appear smooth and can be packed more closely. Mammalian neurofilaments are composed of three proteins of widely different sizes; other filaments are either made up of a single protein or, as in the case of keratin, a number of related peptides of 40,000- to 67,000-mol wt, with keratins from different sources containing several of these polypeptides in different amounts. Like other intermediate filaments, neurofilament proteins are acidic (pIs between 5 and 6) and contain large proportions of aspartyl and glutamyl residues. The amino acid content of the 70K neurofilament component appears most similar to that of other intermediate filaments. (See reference 81 for a summary.) It is also the 70K protein that can be reassembled into smooth-walled 10-nm filaments under assembly conditions similar to those used for other intermediate filaments (29).

Physical-chemical studies with keratins and vimentin filaments have demonstrated an α-helical content as high as 40% and a coiled-coil arrangement of these helical regions, probably during interactions among two or more filament proteins (82). Although similar studies have not been published with mammalian neurofilaments, work with *Myxicola* has also shown a high content of α-helix as well as the coiled-coil configuration.

Neurofilaments are immunologically distinct from other intermediate filaments, but the description of a monoclonal antiserum that cross-reacts with several filaments, including mammalian and invertebrate neurofilaments (61), implies some common amino acid sequences.

REFERENCES

1. Autilio-Gambetti, L., M. E. Velasco, J. Sipple, and P. Gambetti. 1981. Immuno-chemical characterization of antisera to rat neurofilament subunits. *J. Neurochem. 37*:1260-1265.

2. Bear, R. S., F. O. Schmitt, and J. Z. Young. 1937. Investigations on the protein constituents of nerve axoplasm. *Proc. Roy. Soc. Biochem. 123*:520-529.

3. Bell, C. W. 1977. Hydrodynamic properties of neurofilaments. *J. Physiol. 266*:83-84.

4. Berkowitz, S. A., J. Katagiri, H. K. Binder, and R. C. Williams, Jr. 1977. Separation and characterization of microtubule proteins from calf brain. *Biochemistry 16*:5610-5617.

5. Brown, B. A., R. A. Nixon, P. Strocchi, and C. A. Marotta. 1981. Characterization and comparison of neurofilament proteins from rat and mouse CNS. *J. Neurochem. 36*:143-153.

6. Chiu, F.-C., B. Korey, J. E. Goldman, and W. T. Norton. 1979. Peptide analysis of intermediate filament proteins. *Proc. Int. Soc. Neurochem. 7*:275.

7. Chiu, F.-C., B. Korey, and W. T. Norton. 1980. Intermediate filaments from bovine, rat and human CNS: Mapping analysis of the major proteins. *J. Neurochem. 34*:1149-1159.

8. Chiu, F.-C., and W. T. Norton. 1982. Bulk preparation of CNS cytoskeleton and the separation of individual neurofilament proteins by gel filtration: Dye-binding characteristics and amino acid composition. *J. Neurochem. 39*:1252-1260.

9. Chiu, F.-C., W. T. Norton, and K. L. Fields. 1981. The cytoskeleton or primary astrocytes in culture contains actin, glial fibrillary acidic protein, and the fibroblast-type filament protein, vimentin. *J. Neurochem. 37*:147-155.

10. Czosnek, H., and D. Soifer, 1980. Comparison of the proteins of 10 nm filaments from rabbit sciatic nerve and spinal cord by electrophoresis in two dimensions. *FEBS Lett. 117*:175-178.

11. Czosnek, H., D. Soifer, K. Mack, and H. M. Wisinewski. 1981. Similarity of neuro-filament proteins from different parts of the rabbit nervous system. *Brain Res. 216*:387-398.

12. Czosnek, H., D. Soifer, and H. M. Wisniewski. 1980. Heterogeneity of intermediate filament proteins from rabbit spinal cord. *Neurochem. Res. 5*:777-793.

13. Czosnek, H., D. Soifer, and H. M. Wisniewski. 1980. Studies on the biosynthesis of neurofilament proteins. *J. Cell Biol. 85*:726-734.

14. Davison, P. F., and R. N. Jones. 1981. Filament proteins in central, cranial, and peripheral mammalian nerves. *J. Cell Biol. 88*:67-72.

15. Davison, P. F., and E. W. Taylor. 1960. Physical-chemical studies of proteins of squid nerve axoplasm, with special reference to the axon fibrous protein. *J. Gen Physiol. 43*:801-823.

16. Davison, P. F., and B. Winslow. 1974. The protein subunit of calf brain neurofila-ment. *J. Neurobiol. 5*:119-133.

17. Day, W. A., and D. S. Gilbert. 1972. X-ray diffraction pattern of axoplasm. *Biochim. Biophys. Acta 285*:503-506.

18. Deffner, G. G. J., and R. E. Hafter. 1959. Chemical investigations of the giant nerve fibers of the squid: I. Fractionation of dialyzable constituents of axoplasm and quantitative determination of the free amino acids. *Biochim. Biophys. Acta 32*:362-374.

19. Delacourte, A., G. Filliatreau, F. Boutteau, G. Biserte, and J. Schrevel. 1980. Study of the 10 nm-filament fraction isolated during the standard microtubule preparation. *Biochem. J. 191*:543-546.

20. Delacourte, A., M.-T. Plancot, K.-K. Han, H. Hildebrand, and G. Biserte. 1977. Investigation of tubulin fibers formed during microtubule polymerization cycles. *FEBS Lett. 77*:41-46.

21. DeVries, G. H., L. F. Eng., D. L. Lewis, and M. G. Hadfield. 1976. The protein composition of bovine myelin-free axons. *Biochim. Biophys. Acta 439*:133-145.

22. DeVries, G. H., W. T. Norton, and C. S. Raine. 1972. Axons: Isolation from mammalian central nervous system. *Science* 175:1370-1372.

23. Eagles, P., C. W. Bell, A. Maggs, C. Wais, and D. S. Gilbert. 1978. Neurofilament structure and its enzymatic modification. *In*: The Cytoskeletal and Contractile Networks of Non-Muscle Cells. R. Goldman, editor. Cold Spring Harbor Laboratory, Cold Spring Harbor, N.Y., P. 47.

24. Eagles, P. A. M., D. S. Gilbert, and A. Maggs. 1981. The location of phosphorylation sites and Ca^{2+}-dependent proteolytic cleavage sites on the major neurofilament polypeptides from *Myxicola infundibulum*. *Biochem. J.* 199:101-111.

25. Eagles, P. A. M., D. S. Gilbert, and A. Maggs. 1981. The polypeptide composition of axoplasm and of neurofilaments from the marine worm *Myxicola infundibulum*. *Biochem. J.* 199:89-100.

26. Eng, L. F., R. J. Lasek, J. W. Bigbee, and D. L. Eng. 1980. Immunocytological analyses of 10 nm intermediate filaments in the nervous system of *Myxicola*. *J. Histochem. Cytochem.* 28:1312-1318.

27. Eng, L. F., J. J. Vanderhaeghen, A. Bignami, and B. Gerstl. 1971. An acidic protein isolated from fibrous astrocytes. *Brain Res.* 28:351-354.

28. Gaskin, F., S. B. Kramer, C. R. Cantor, R. Adelstein, and M. L. Shelanski. 1974. A dynein-like protein associated with neurotubules. *FEBS Lett.* 40:281-286.

29. Geisler, N., and K. Weber. 1981. Self assembly *in vitro* of the 68,000 molecular weight component of the mammalian neurofilament triplet proteins into intermediate-sized filaments. *J. Mol. Biol.* 151:565-571.

30. Gilbert, D. S. 1975. Axoplasm architecture and physical properties as seen in the *Myxicola* giant axon. *J. Physiol.* 253:257-301.

31. Gilbert, D. S., B. J. Newby, and B. H. Anderton. 1975. Neurofilament disguise, destruction, and discipline. *Nature* 256:586-589.

32. Glicksman, M. A., and M. Willard. 1982. Separation of neurofilaments from ATPase activity by precipitation with anti-neurofilament antibodies. *J. Neurochem.* 38:1774-1776.

33. Goldman, J. E., H. H. Schaumburg, and W. T. Norton. 1978. Isolation and characterization of glial filaments from human brain. *J. Cell Biol.* 78:426-440.

34. Griffin, J. W., P. N. Hoffman, A. W. Clark, P. T. Carroll, and D. L. Price. 1978. Slow axonal transport of neurofilament proteins: Impairment by β-β'-iminodipropionitrile administration. *Science* 202:633-635.

35. Hodgkin, A. L., and B. Katz, 1949. The effect of calcium on the axoplasm of giant nerve fibers. *J. Exp. Biol.* 26:292-294.

36. Hoffman, P. N., and R. J. Lasek. 1975. The slow component of axonal transport: Identification of major structural polypeptides of the axon and their generality among mammalian neurons. *J. Cell Biol.* 66:351-366.

37. Hogue-Angeletti, R. A., H.-L. Wu, and W. W. Schlaepfer. 1982. Preparative separation and amino acid composition of neurofilament triplet proteins. *J. Neurochem.* 38:116-120.

38. Huneeus, F. C., and P. F. Davison, 1970. Fibrillar proteins from squid axons: I. Neurofilament protein. *J. Mol. Biol.* 52:415-428.

39. Hynes, R. O., and A. T. Destree. 1978. 10 nm filaments in normal and transformed cells. *Cell* 13:151-163.

40. Julien, J.-P., and W. E. Myshynski. 1981. A comparison of *in vitro*- and *in vivo*-phosphorylated neurofilament polypeptides. *J. Neurochem.* 37:1579-1585.

41. Krishnan, N., I. R. Kaiserman-Abramof, and R. J. Lasek. 1979. Helical substructure of neurofilaments isolated from *Myxicola* and squid giant axons. *J. Cell Biol.* 82:323-335.

42. Lasek, R. J., and M. M. Black. 1977. How do axons stop growing? Some clues from the metabolism of the proteins in the slow component of axonal transport. *In*: Mechanisms, Regulation and Special Functions of Protein Synthesis in the Brain. S. Roberts, A. Lajtha, and W. H. Gispen, editors. Elsevier/North Holland Biomedical Press, Amsterdam, pp. 161-169.

43. Lasek, R. J., and P. N. Hoffman. 1976. The neuronal cytoskeleton, axonal transport, and axonal growth. *In*: Cell Motility. Vol. 3, Microtubules and Related Proteins. R. Goldman, T. Pollard, and J. Rosenbaum, editors. Cold Spring Harbor Laboratory, Cold Spring Harbor, N.Y., pp. 1021-1049.

44. Lasek, R. J., N. Krishnan, and I. R. Kaiserman-Abramof. 1979. Identification of the subunit proteins of 10 nm neurofilaments isolated from axoplasm of squid and *Myxicola* giant axons. *J. Cell Biol. 82*:336-346.

45. Lazarides, E. 1980. Intermediate filaments as mechanical integrators of cellular space. *Nature 283*:249-256.

46. Leterrier, J.-F., R. K. H. Liem, and M. L. Shelanski. 1981. Preferential phosphorylation of 150,000 molecular weight component of neurofilaments by a cyclic AMP-dependent, microtubule-associated protein kinase. *J. Cell Biol. 90*:755-760.

47. Liem, R. 1982. Simultaneous separation and purification of neurofilament and glial filament proteins from brain. *J. Neurochem. 38*:142-150.

48. Liem, R. K. H., and S. B. Hutchison. 1982. Purification and individual components of the neurofilament triplet; Filament assembly from the 70,000 dalton subunit. *Biochemistry 21*:3221-3226.

49. Liem, R. K. H., S.-H. Yen, G. D. Salomon, and M. L. Shelanski. 1978. Intermediate filaments in nervous tissue. *J. Cell Biol. 79*:637-645.

50. Maxfield, M., and R. W. Hartley. 1957. Dissociation of the fibrous protein of nerve. *Biochim. Biophys. Acta 24*:83-87.

51. Metuzals, J. 1969. Configuration of a filamentous network in the axoplasm of the squid (*Loligo pealii*) giant nerve fiber. *J. Cell Biol. 43*:480-505.

52. Metuzals, J., and C. S. Izzard. 1969. Spatial patterns of threadlike elements in the axoplasm of the giant nerve fibers of the squid (*Loligo pealii*) as disclosed by differential interference microscopy and by electron microscopy. *J. Cell Biol. 43*:456-479.

53. Moon, H. M., T. Wisniewski, P. Merz, J. DeMartini, and H. M. Wisniewski. 1981. Partial purification of neurofilament subunits from bovine brains and studies on neurofilament assembly. *J. Cell Biol. 89*:560-567.

54. Mori, H., and M. Kurokawa. 1980. Morphological and biochemical characterization of neurofilaments isolated from the rat peripheral nerve. *Biomed. Res. 1*:24-31.

55. Nakayama, T. 1981. Synthesis of cytoskeletal proteins in bulk-isolated neuronal perikarya. *J. Neurochem. 36*:1398-1405.

56. Norton, W. T., and J. E. Goldman. 1980. Neurofilaments. *In*: Proteins of the Nervous system. R. A. Bradshaw and D. M. Schneider, editors. Raven Press, New York, pp. 301-329.

57. Norton, W. T., and J. M. Turnbull. 1970. The isolation and lipid composition of a myelin-free axon-enriched fraction from the CNS. *Fed. Proc. 29*:472 (abs.)

58. Pant, H., and H. Gainer. 1980. Properties of a calcium-activated protease in squid axoplasm which selectively degrades neurofilament proteins. *J. Neurobiol. 11*:1-12.

59. Pant, H., S. Terakawa, and H. Gainer. 1979. A calcium activated protease in squid axoplasm. *J. Neurochem. 32*:99-102.

60. Pant, H., T. Yoshioka, I. Tasaki, and H. Gainer. 1979. Divalent cation dependent phosphorylation of proteins in squid giant axon. *Brain Res. 162*:303-313.

61. Pruss, R. M., R. Mirsky, M. C. Raff, R. Thorpe, A. J. Dowding, and B. H. Anderton. 1981. All classes of intermediate filaments share a common antigenic determinant defined by a monoclonal antibody. *Cell 27*:419-428.

62. Runge, M. S., H. W. Detrich, and R. C. Williams, Jr. 1979. Identification of the major 68,000-Dalton protein of microtubule preparations as a 10-nm filament protein and its effects on microtubule assembly *in vitro*. *Biochemistry 18*:1689-1698.

63. Runge, M. S., P. B. Hewgley, D. Puett, and R. C. Williams, Jr. 1979. Cyclic nucleotide phosphodiesterase activity in 10-nm filaments and microtubule preparations from bovine brain. *Proc. Natl. Acad. Sci. USA 76*:2561-2565.

64. Runge, M. S., T. M. Laue, D. A. Yphantis, M. R. Lifsics, A. Saito, M. Altin, K. Reinke, and R. C. Williams, Jr. 1981. ATP-induced formation of an associated complex between microtubules and neurofilaments. *Proc. Natl. Acad. Sci. USA.* *78*:1431-1435.

65. Runge, M. S., M. Raafat El-Maghrabi, T. H. Claus, S. J. Pilkis, and R. C. Williams, Jr. 1981. A MAP-2-stimulated protein kinase activity associated with neurofilaments. *Biochemistry* *20*:175-180.

66. Runge, M. S., W. W. Schlaepfer, and R. C. Williams, Jr. 1981. Isolation and characterization of neurofilaments from mammalian brain. *Biochemistry* *20*:170-175.

67. Schachner, M., C. Smith, and G. Schoonmaker. 1978. Immunological distinction between neurofilament and glial fibrillary acidic proteins by mouse antisera and their immunohistological characterization. *Devel. Neurosci.* *1*:1-14.

68. Schlaepfer, W. W. 1977. Studies on the isolation and substructure of mammalian neurofilament. *J. Ultrastruct. Res., 61*:149-157.

69. Schlaepfer, W. W. and L. A. Freeman. 1978. Neurofilament proteins of rat peripheral nerve and spinal cord. *J. Cell Biol.* *78*:653-662.

70. Schlaepfer, W. W., V. Lee, and H.-L. Wu. 1981. Assessment of immunological properties of neurofilament triplet proteins. *Brain Res.* *226*:259-272.

71. Schook, W. J., and W. T. Norton. 1976. Neurofilaments account for the lipid in myelin-free axons. *Brain Res.* *118*:517-522.

72. Selkoe, D. J., R. K. H. Liem, S.-H. Yen, and M. L. Shelanski. 1979. Biochemical and immunological characterization of neurofilaments in experimental neurofibrillary degeneration induced by aluminum. *Brain Res.* *163*:235-252.

73. Shecket, G., and R. J. Lasek. 1979. Phosphorylation of neurofilaments from mammalian peripheral nerve. *Trans. Am. Soc. Neurochem.* *10*:140 (abs.).

74. Schecket, G., and R. J. Lasek. 1980. Preparation of neurofilament protein from guinea pig peripheral nerve and spinal cord. *J. Neurochem.* *35*:1335-1344.

75. Shelanski, M. L., S. Albert, G. H. DeVries, and W. T. Norton. 1971. Isolation of filaments from brain. *Science 174*:1242-1245.

76. Shelanski, M. L., J.-F. Leterrier, and R. K. H. Liem. Evidence for interactions between neurofilaments and microtubules. *Neurosciences Res. Prog. Bull.* *19*:32-43.

77. Shelanski, M. L., and R. K. H. Liem. 1979. Neurofilaments. *J. Neurochem.* *33*:5-13.

78. Sloboda, R. D., S. A. Rudolph, J. L. Rosenbaum, and P. Greengard. 1975. Cyclic AMP-dependent endogenous phosphorylation of a microtubule-associated protein. *Proc. Natl. Acad Sci. USA 72*:177-181.

79. Small, J. V., and A. Sobieszek. 1977. Studies on the function and composition of the 10 nm (100 Å) filaments of vertebrate smooth muscle. *J. Cell Sci.* *23*:243-268.

80. Soifer, D., K. Iqbal, H. Czosnek, J. DeMartini, J. Sturman, and H. M. Wisniewski. 1981. The loss of neuron-specific proteins during the course of Wallerian degeneration of optic and sciatic nerve. *J. Neurosci.* *1*:461-470.

81. Starger, J. M., W. E. Brown, A. E. Goldman, and R. D. Goldman. 1978. Biochemical and immunological analysis of rapidly purified 10 nm filaments from baby hamster kidney (BHK-21) cells. *J. Cell Biol.* *78*:93-109.

82. Steinert, P. M., S. B. Zimmerman, J. M. Starger, and R. D. Goldman. 1978. Ten nanometer filaments of hamster BHK-21 cells and epidermal keratin filaments have similar structures. *Proc. Natl. Acad. Sci. USA 75*:6098-6101.

83. Tapscott, S. J., G. S. Bennett, and H. Holtzer. 1981. Neuronal precursor cells in the chick neural tube express neurofilament proteins. *Nature 292*:836-838.

84. Tapscott, S. J., G. S. Bennett, Y. Toyama, F. Kleinbart, and H. Holtzer. 1981. Intermediate filament proteins in the developing chick spinal cord. *Devel. Biol.* *86*:40-54.

85. Thorpe, R., A. Delacourte, M. Ayers, C. Bullock, and B. H. Anderton. 1979. The polypeptides of isolated brain 10 nm filaments and their association with polymerized tubulin. *Biochem. J.* *181*:275-284.

56   Chiu, Goldman, and Norton

86. Weiss, P., and H. B. Hiscoe. 1948. Experiments on the mechanism of nerve growth. *J. Exp. Zool.* 107:315-395.

87. Willard, M., and C. Simon. 1981. Antibody decoration of neurofilaments. *J. Cell Biol.* 89:198-205.

88. Wisniewski, H. M., Shelanski, M. L., and Terry, R. L. 1968. Effects of mitotic spindle inhibitors on neurotubules and neurofilaments in anterior horn cells. *J. Cell Biol.* 38:224-229.

89. Wuerker, R. B. 1970. Neurofilaments and glial filaments. *Tiss. Cell* 2:1-9.

90. Yen, S.-H., D. Dahl, M. Schachner, and M. L. Shelanski. 1976. Biochemistry of the filaments of brain. *Proc. Natl. Acad. Sci. USA* 73:533.

91. Yen, S.-H., and K. L. Fields. 1981. Antibodies to neurofilament, glial filament and fibroblast intermediate filament proteins bind to different cell types of the nervous system. *J. Cell Biol.* 88:115-126.

92. Zackroff, R. V., and R. D. Goldman. 1980. *In vitro* reassembly of squid brain intermediate filaments (neurofilaments): Purification by assembly-disassembly. *Science* 208:1152-1155.

Chapter 3

Neurofilaments of Mammalian Peripheral Nerve

William W. Schlaepfer

Neurofilaments and the Cytoskeleton of Peripheral Nerve

Peripheral nerves are composed of neuronal processes or axons that extend for extraordinary distances from neuronal perikarya. These axonal processes are supported by a cytoskeleton that occupies almost all of the axonal content. The abundance of this axonal cytoskeleton, its accessibility, as well as the parallel alignment and proximodistal orientation of nerve fibers provide advantages to studies of the cytoskeleton in peripheral nerve.

Neurofilaments represent the most conspicuous component of the cytoskeleton in peripheral nerve. Under electron microscopy, it is evident that the cytoskeleton consists of neurofilaments and microtubules arrayed along the longitudinal axis of the nerve fibers (115, 116). Actin and myosin are also evident in axons when assessed by biochemical and immunochemical methods (7,46,53,99,111,112), but the nature of their localizations is not revealed by defined morphological structures. The nature of interactions among cytoskeletal components is unknown but is probably complex. Some insight into these relationships may be reflected in the relative distribution of cytoskeletal components within and among different nerve fibers.

The distributional relationship between neurofilaments and microtubules has been most clearly elucidated in the peripheral nerve. For example, microtubules are the predominant cytoskeletal component in thin unmyelinated axons, their density diminishing in larger axons at a rate that is roughly proportional to the inverse of axonal size (115,116). On the average, there are 50 to 100 microtubules per

57

square micrometer in unmyelinated axons but only 10 to 20 micro-tubules per square micrometer in large, myelinated axons (6). Micro-tubules are usually distributed relatively uniformly in small axons. In large axons, however, they tend to be clustered, sometimes in close association with axonal mitochondria or vesicular structures.

Neurofilaments, in contrast to microtubules, are most conspicuous in large myelinated axons. In the largest fibers, neurofilaments are distributed homogeneously, nearly filling the axoplasmic compart-ment (115). In smaller fibers, the number of neurofilaments dimin-ishes, and yet their density remains about the same as they become admixed with increasing numbers of microtubules (30). It is of inter-est that the same number of neurofilaments per cross-sectional area is found in small and large unmyelinated axons (6). Furthermore, the same density of neurofilaments is present in unmyelinated and my-elinated axons (6). Neurofilaments of unmyelinated fibers, however, are often clustered in their distribution within the axon.

It is of interest that the relative distributions of neurofilaments and microtubules are not constant along the course of individual axons. In particular, different cytoskeletal arrangements are noted in the ini-tial segment of the axon and in the paranodal axoplasm of myelinated nerve fibers. The initial segment is situated immediately proximal to the first myelin sheath and is characterized by an undercoating of the axolemma (69), even on primary sensory neurons that lack synaptic input (120). Axoplasm of the initial segments contains abundant neurofilaments with an increased number of microtubules, the latter often occurring in clusters with interconnecting elements and referred to as fasciculated microtubules. A similar undercoating of axolemma occurs at nodal areas of myelinated fibers (26,76). Likewise, nodal axoplasm is characterized by an increase in the density of micro-tubules with a clustering similar to that of the initial axonal segment (6). Nodal axoplasm represents a constricted region of axon with cross-sectional area reduced to 50% or more (6). Neurofilaments are diminished in nodal axoplasm but retain a density similar to that of internodal axoplasm (6).

An asymmetry of cytoskeletal elements has been noted between the proximal and distal bifurcation of the primary sensory neurons (36). In particular, a relative paucity of microtubules occurs in the posterior spinal nerve roots when compared with ventral nerve root fibers or peripheral sensory fibers of the same size (103,121). The relative deficit of microtubules may represent an increase of unpoly-merized subunit since the relative amounts of neurofilament proteins to tubulin is 1.05 in dorsal root but only 0.65 in peripheral nerve (54). The diminished numbers of microtubules in dorsal roots are al-so associated with decreased rates of slow axonal transport, affecting both SCa and SCb components of axonal flow (51,66). It has been speculated that different compositions of proximal and distal axonal cytoskeletons of the primary sensory neurons reflect their separate origins within the neuronal cell body (54).

The preponderant occurrence of neurofilaments in large axons is also reflected by the increased numbers of neurofilaments in the peri-karya of large neurons (2,108). For example, neurofilaments are much more conspicuous in the large, pale neurons than in the small, dark neurons of the sensory ganglia. In the large neurons, parallel arrays of neurofilaments and microtubles course between aggregates of rough endoplasmic reticulum, thereby compartmentalizing the latter within the perikaryon and accounting for its discontinuous staining as Nissl substance. Filamentous arrays funnel into the axon hillock, where their orientation becomes aligned with that of the axon. The axon hillock is said to contain fasciculated as well as an increased number of microtubules, similar to that of the adjoining initial segment of axon (108).

Neurofilaments are poorly defined and sparsely scattered in the perikarya of small, dark neurons of primary sensory ganglia. Their paucity is reflected by the limited immunofluorescence of these cells when reacted with neurofilament antisera (92). More immunofluores-cence is seen in unmyelinated fibers, indicating the general prepon-derance of neurofilaments in neurites than in the perikarya of the same cell. Moreover, the demonstration of neurofilament immuno-fluorescence in unmyelinated fibers provides evidence that small, dark sensory neurons are neurofilament-producing cells that operate at reduced levels of neurofilament production. Likewise, neurons of the sympathetic trunk are neurofilament-poor cells in which neuro-filament immunofluorescence is more readily demonstrated in their neurites than in their cell bodies (92).

Development of Peripheral Nerves and Neurofilaments

Neurofilaments and the axonal cytoskeleton undoubtedly play a critical role during nerve development. Studies on chick embryos in-dicate that neurofilament proteins appear initially in postmitotic neuroblasts at about the time of the initial axon formation (106). Younger proliferating neuroblasts contain vimentin-type intermediate filaments, which are replaced by neurofilaments in postmitotic cells (106). Similar distributions of neurofilament proteins are noted through the use of antisera to the largest and smallest neurofilament triplet proteins of the chick (106), suggesting that synthesis of these different neurofilament proteins is initiated at the same time in the chick embryo. Studies on neurofilament synthesis in cell-free systems indicate that each of the neurofilament triplet proteins is synthesized as an independent event (15).

Maturation of neuroblasts can also be measured by their ability to express neurofilament proteins when explanted and grown *in vitro*. This ability is initially acquired by rat neural tissues after 12 days of enbryonic growth (74). Neuroblasts of younger embryos proliferate *in vitro* but do not produce neurofilaments or synthesize neurofilament proteins even after prolonged periods of incubation. Accordingly, there appears to be a maturation factor that is operative during proliferation of neuroblasts *in vitro* and that is instrumental in turning on neurofilament protein synthesis. It is possible that this factor results from the release of inhibitory influences.

Morphological studies on the development of the dorsal root ganglia indicate that intermediate filaments are scattered in the cytoplasm of neuroblasts prior to the formation of neuritic processes (107). Similar filaments are seen in differentiating motor neuroblasts in chick neural tube (60). Whether such filaments represent neurofilaments or vimentin filaments cannot be determined by morphological criteria. The developing axonal cytoskeleton of dorsal root neuroblasts also contains intermediate filaments that are indistinguishable from neurofilaments (107). Fewer intermediate filaments are present in developing neurites of small, dark neurons than in large, light neurons, especially in the initial segments of the latter cells. It is noteworthy that intermediate filaments are much fewer in number than microtubules in the developing neurite. Similar observations have been noted in developing neurites *in vitro* (10). Absence of neurofilaments in the early stages of neurite development and their appearance at a latter stage of development is reported in rat optic nerve (70) and in developing spinal cord of kittens (102) and monkeys (8).

Morphology of Neurofilaments

The abundance of neurofilaments and their longitudinal alignment in peripheral nerve have facilitated a detailed analysis of their ultrastructural appearance (48,115,116,117). Cross-sectioned neurofilaments are roughly circular, measuring 8 to 11 nm in diameter. End-on profiles of neurofilaments reveal an electron-lucent core surrounded by walls with a beaded appearance. This configuration is consistent with helically arrayed subunits.

Although the morphology of individual neurofilaments is not distinguished from other intermediate filaments, the spatial orientation of neurofilaments is distinctive (115). In particular, neurofilaments are arrayed in an orderly parallel arrangement in which minimal distances of 25 to 30 nm are maintained between adjacent

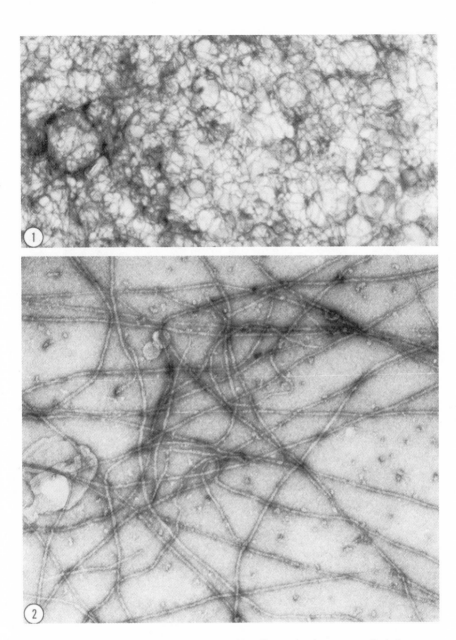

Figure 3-1. Overlapping neurofilaments comprising the predominant organelle in low-speed supernatant fractions from osmotically shocked rat peripheral nerve. (Preparation was negatively stained with uranyl acetate after fixation with 5% formalin. (Magnified ×5,000.) (From reference 83. Used with permission.)

Figure 3-2. Individual neurofilaments coursing as discrete cylindrical structures with relatively smooth external surfaces. The widths of neurofilamentous profiles generally vary from 80 to 100 Å, but thinner segments occur, often near the termination of individual neurofilaments. (Fixed and stained as in Figure 3-1. Magnified × 100,000.) (From reference 83. Used with permission.)

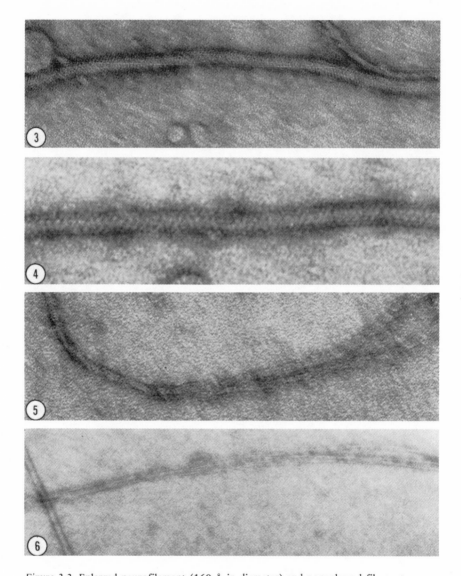

Figure 3-3. Enlarged neurofilament (160 Å in diameter) and nonenlarged filament (upper right) following a 1-hour incubation in urea (500 mM) and sodium chloride (75 mM). The enlarged filament reveals a herringbone substructure indicative of a helical arrangement of constituent protofilaments. (Fixed and stained as in Figure 3-1. Magnified X 250,000.) (From reference 83. Used with permission.)

Figure 3-4. Herringbone substructural pattern in neurofilament enlarged to 210 Å in diameter following a 1-hour incubation in urea (500 mM) and sodium chloride (75 mM). Crisscrossing protofilaments measure 20 to 30 Å in width. (Fixed and stained as in Figure 3-1. Magnified X 335,000.) (From reference 83. Used with permission.)

Figure 3-5. Splaying of neurofilamentous terminus into paired filamentous subunits

filaments, even during conditions of neurofilament accumulations. It is not known whether the regular spacing and parallel alignment of neurofilaments is associated with the lateral sidearms that project perpendicularly into the axoplasm (115,116,117). In tissues impregnated with heavy metals, both neurofilaments and their lateral sidearms as well as microtubules have been visualized as part of a three-dimensional cytoplasmic network (57,58,61,62). A similar axoplasmic latticework of interconnecting thin strands ranging in size from 4 to 10 nm has been observed by high-voltage electron microscopy in tissues that have been subjected to critical-point-drying in order to minimize artifact (27,114). (See Chapter 1.)

Intact neurofilaments have been separated from rat peripheral nerve so that their profiles can be visualized by negative staining techniques (83), as illustrated in Figures 3-1 and 3-2. Individual filaments appear as discrete, unbranching, linear structures that extend for distances of at least 10 μm. Their course is often curved, thereby indicating an inherent flexibility in their construction. Neurofilament profiles are those of smooth, cylindrical structures without evidence of lateral sidearms. It is unclear whether the lateral sidearms seen by electron microscopy on filaments *in situ* are not preserved, become detached, or collapse along the axial core during the separation and isolation of filaments. The lateral margins of the filaments are parallel and subtend a cylindrical structure of 8 to 11 nm diameter. Thinner segments are encountered, often near tapered ends of filaments. Neurofilaments from invertebrate giant axons have been noted to become frayed or untwisted at their termini, thereby revealing multiple 1.5- to 5-nm protofilaments (33,52).

Substructural features of mammalian neurofilaments can be visualized in filaments whose intermolecular bonds are loosened by exposure to urea prior to fixation (83). This treatment results in expansion of the axial core (Figures 3-3 and 3-4) and visualization of protofilamentous elements (Figures 3-5 and 3-6). Urea-enlarged neurofilaments measure up to 26 nm in diameter and display a herringbone substructure (Figures 3-3 and 3-4), suggesting a helical or spiral arrangement of linear subunits. The crisscrossing linear subunits measure 2.0 to 2.5 nm in width and form an angle of $28° \pm 4°$ with the horizontal plane. Occasionally, urea-treated neurofilaments are

following a 1-hour exposure to urea (500 mM) and sodium chloride (75 mM). (Fixed and stained as in Figure 3-1. Magnified X 275,000.) (From reference 83. Used with permission.)

Figure 3-6. Neurofilamentous profile revealing multiple parallel protofilamentous components following a 1-hour incubation in urea (500 mM) and sodium chloride (75 mM). Intact neurofilament is present at lower left. (Fixed and stained as in Figure 3-1. Magnified X 100,000.) (From reference 83. Used with permission.)

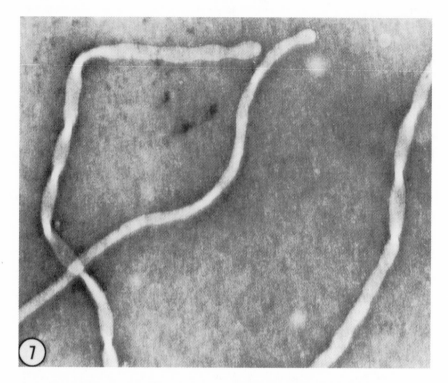

Figure 3-7. Multiple twisting deformations occurring along longitudinal axes of enlarged neurofilaments after 21-day incubation in 0.1-M potassium chloride. Successive twisting deformations occur at 800 to 1000-Å intervals. Filament diameters measure 100 to 150 Å in areas of twists and approximately 250 Å in the intervening segments. From reference 84. Reproduced from the *Journal of Neuropathology and Experimental Neurology*, 1978, vol. 38, pp. 244-254.) (Formalin-fixed preparation, negatively stained with uranyl acetate. Magnified X 200,000.)

splayed apart into two (Figure 3-5) or more (Figure 3-6) components. Separated protofilaments measure 2 to 3 nm in width. Invertebrate neurofilaments are also composed of 2- to 3-nm protofilaments that interdigitate in a twisting or spiraling construction (52). It is suggested that the helical architecture inherent in neurofilaments accounts for many physical properties of axoplasm, including the spiraling of myelin around the axon (31,32).

The inherent helical substructure of neurofilaments is evidence at several levels of organization. X-ray-diffraction analyses of neurofilaments reveal patterns indicative of considerable coiled-coil α-helical configuration in the peptide sequences of neurofilament subunits (20). Similar α-helical configuration is estimated to comprise from 40% to 80% of the amino acid sequence in subunits of other intermediate filaments and is believed to represent a common trait of all intermediate filaments (104). Extensive coiled-coil α-helix in neurofilament subunits could be expected to produce torque along the axial cord of the assembled neurofilament, especially if some bonds become preferentially loosened. The presence of torque in neurofilaments is, in fact, evidenced by the twisting deformations that occur along intact neurofilaments when subjected to time-dependent decay *in vitro* (84), as illustrated in Figure 3-7. Whereas urea treatment of neurofilaments tends to uncover intact protofilaments (Figure 3-2, 3-3, 3-4, 3-5, and 3-6), the time-dependent breakdown of neurofilaments does not lead to separation of protofilaments or their visualization within the assembled filament structure. Instead, enlargement of neurofilament profiles is accompanied by periodic twists along the neurofilament axis. Similar twisted filamentous configurations occur in human neurons and neurites associated with neurofibrillary plaques and tangles (50,113). It is not known whether the twisted filaments of neurofibrillary pathology also arise from alterations of neuronal cytoskeletal proteins. (See Chapter 7.)

Isolation of intact neurofilaments from peripheral nerve has also enabled the identification of conditions that lead to filament disassembly (85), as outlined in Table 3-1. Of particular practical interest is the instability of neurofilaments in hypotonic media in contrast to previously reported stability of invertebrate neurofilaments under similar conditions (41). This new insight provided an explanation for the initial confusion regarding the identity of neurofilament protein and its relationship to glial fibrillary acidic protein. Initial isolation of mammalian neurofilament protein utilized a prolonged period of hypotonic shock to separate neurofilaments in axons of central myelinated fibers from their myelin sheaths (23,101). It was not realized that this treatment disrupted neurofilaments and led to enrichment of glial filaments rather than neurofilaments. Consequently, there remain many reports in the early literature on neurofilament biochemistry that assume that neurofilaments are composed of a 50,000- to 54,000-mol wt subunit and possess marked similarity with glial fibrillary acidic protein (GFA) (18,19,22,44,97,98,118).

Neurofilament Proteins and Peripheral Nerve

Peripheral nerve has served as an important tissue for the identification and characterization of neurofilament protein subunits. Studies of peripheral nerve proteins refuted the initial theory that neurofilaments and glial filaments shared the same protein subunits. In particular, the absence of GFA protein in peripheral nerve (1,12,40,49, 55,59,64,67,88,93,94,100) indicated that neurofilament protein subunits of peripheral nerve could not be represented by the prominent 50,000- to 54,000-mol wt GFA protein band found in SDS gels of CNS white-matter tissue (illustrated in Figure 3-8). This GFA protein band can be identified with glial filaments that extend into the nerve-root entry zone, but it is not present in more-distal segments of the same peripheral nerve fibers (90).

The absence of GFA protein facilitated the use of peripheral nerve

Table 3-1
Effects of Incubational Media on Preservation of Neurofilaments

Media	Concentration mM/liter	Demonstration of Intact Neurofilaments*				
		1h[†]	4h[†]	8h[†]	16h[†]	24h[§]
H$_2$O	–	1+	0	0	0	0
2-mercaptoethanol	50	2+	1+	1+	1+	0
Colchicine	1	1+	1+	1+	0	0
Vinblastine	1	1+	1+	1+	1+	0
NaCl (or KCl)	10	2+	1+	1+	1+	1+
NaCl (or KCl)	20	2+	2+	2+	1+	1+
NaCl (or KCl)	100	2+	2+	2+	2+	2+
Urea (or guanidine HCl)	500	0	0	0	0	0
Urea	500	2+	2+	2+	1+	1+
NaCl	100	2+	2+	2+	1+	1+
NaPO$_4$-buffer, pH 5.2	10	2+	1+	1+	0	0
NaCl	100	2+	1+	1+	0	0
NaPO$_4$-buffer, pH 5.7	10	2+	2+	2+	1+	1+
NaCl	100	2+	2+	2+	1+	1+
NaPO$_4$-buffer, pH 6.2	10	2+	2+	2+	2+	2+
NaCl	100	2+	2+	2+	2+	2+
NaPO$_4$-buffer pH 6.7	10	2+	2+	2+	2+	2+
NaCl	100	2+	2+	2+	2+	2+
NaPO$_4$-buffer, pH 7.2	10	2+	2+	1+	1+	0
NaCl	100	2+	2+	1+	1+	0
MES-buffer, pH 6.2	10	2+	2+	2+	2+	2+
NaCl	100	2+	2+	2+	2+	2+
Tris-buffer, pH 7.2	10	2+	2+	1+	1+	1+
NaCl	100	2+	2+	1+	1+	1+
NaPO$_4$-buffer, pH 6.2	10	2+	2+	2+	1+	1+
MES-buffer, pH 6.2	10	2+	2+	2+	1+	1+

Source: Reference 85. Reproduced from *The Journal of Cell Biology*, 1978, vol. 76, pp. 50-56, by copyright permission of the Rockefeller University Press.

*Estimation of neurofilament structural integrity based upon the ease with which intact neurofilaments could be visualized and the prevalence of their breakdown products in negatively stained preparations.

[†]Examinations conducted on neurofilament-rich supernates dialyzed versus media.

[§]Examinations conducted on fixed sediment of samples after centrifugation.

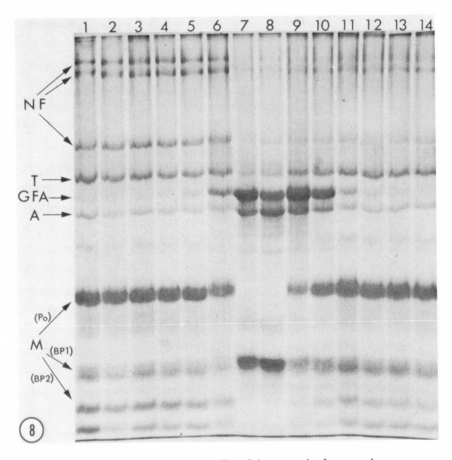

Figure 3-8. Comparative electrophoretic profiles of tissue proteins from anterior roots (lanes 1-6), lateral funiculus of spinal cord (lane 7), posterior funiculus of spinal cord (lane 8), and posterior roots (lanes 9-14) from 77-year-old-man who died 4 hours prior to autopsy. Samples of anterior roots (lanes 1 through 6) and posterior roots (lanes 9 through 14) represent successive 1-mm segments of roots, as measured from the pial surface. (Abbreviations: neurofilament proteins [NF], tubulin [T], glial fibrillary acidic protein [GFA], actin [A], and myelin proteins [M]. A 1.5-mm slab gel of 12% acrylamide run with Tris-glycine buffer, pH 8.3, with approximately 50 μg protein in each sample.) (From reference 90. Used with permission.)

for the identifications of neurofilament proteins. Studies of axonal proteins and their rates of transport along peripheral nerve led to the recognition of 212,000-, 160,000-, and 68,000-mol wt proteins as major components of axonal flow (40). These three proteins moved with tubulin as the major elements of slow axonal flow. All three proteins moved in synchrony with each other and were suspected of representing a major structural component of the axoplasm. Accordingly, it was proposed that the three proteins represented intrinsic components of neurofilaments, thus giving rise to the concept of neurofilament triplet proteins (40).

Immunochemical studies of peripheral nerve proteins provided initial support for the hypothesis of neurofilament triplet proteins (82). Antibodies raised to a 68,000-mol wt protein from peripheral nerve reacted with surface components on isolated rat neurofilaments (Figures 3-9 and 3-10) and localized to neurofilament structures in rat and human central and peripheral nervous systems (92). Reactivity of antibody with neurofilaments was eliminated by absorption with gel-excised 68,000-mol wt nerve protein but not with rat serum or bovine serum albumin, thereby disproving the suggestion that the reactive 68,000-mol wt protein of peripheral nerve was albumin rather than an intrinsic component of neurofilaments (59).

Biochemical studies on peripheral nerve gave further support to the concept of neurofilament triplet proteins and, in particular, to

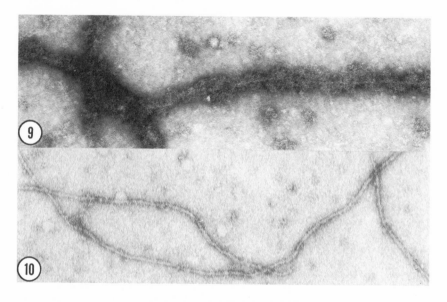

Figure 3-9. Antibody decoration obscuring the profiles of isolated neurofilaments after their exposure to experimental antisera raised to a 68,000-mol wt protein in rat peripheral nerve. Antibody decoration was not altered by absorbing antisera with rat serum or serum albumin. (Magnified ×100,000.) (From reference 82. Reproduced from *The Journal of Cell Biology*, 1977, vol. 74, pp. 226-240, by copyright permission of the Rockefeller University Press.)

Figure 3-10. Neurofilament profiles not obscured after exposure to control antisera under the same conditions as in Figure 3-9. (Magnified × 100,000.) (From reference 82 Reproduced from *The Journal of Cell Biology*, 1977, vol. 74, pp. 226-240, by copyright permission of the Rockefeller University Press.)

the admixture of three proteins in neurofilament structure. Axonal preparations from peripheral nerve that are almost exclusively composed of neurofilaments by electron microscopy (Figure 3-11) display the neurofilament triplet proteins as their major protein constituents on SDS gel electrophoretograms (Figure 3-12). These axonal preparations do not reveal a 50,000- to 54,000-mol wt protein comparable to that found in "neurofilament" preparations from brain tissues, thereby highlighting the discrepancy regarding neurofilament protein subunits found in the central versus peripheral nervous systems. This discrepancy was eventually resolved as several different laboratories demonstrated the ubiquity of neurofilament triplet proteins in the central as well as peripheral nervous systems in mammalian (1,9,11,12,13,14,16,59,72,77,82,88,100,109,110) and avian (28) species.

Morphological Alterations of Neurofilaments by Calcium

Peripheral nerve has served as a useful model for the study of the effects of experimental manipulations on neurofilaments. For example, alterations of neurofilaments *in situ* can be brought about by incubating very small segments of nerve in different solutions prior to fixation and processing for electron microscopy (78,79). Neurofilaments are inherently stable under these conditions and can be readily demonstrated following prolonged periods of incubation in isotonic-neutral buffered solutions (Figure 3-13). Exposure to hypotonic solutions, however, leads to a rapid and progressive loss of neurofilaments (79), thereby providing corroborative evidence that mammalian neurofilaments are unstable under hypotonic conditions (see above).

Disruption of neurofilaments *in situ* was also shown to occur during experimental incubation of peripheral nerve with calcium (78). By electron microscopy, the calcium effects were characterized by a granular disintegration of axonal neurofilaments and microtubules (Figure 3-14). Calcium disruption of neurofilaments was initially noted during incubations in Ringer's solution but could be reproduced by exposure to 1-mM calcium in isotonic saline or in isotonic solutions neutralized with phosphate, Tris, or cacodylate buffers. The calcium-induced changes of neurofilaments could be prevented by chelating calcium with equimolar concentrations of EDTA and were not simulated by magnesium, aluminum, or lead or by the raised or lowered pH of the incubation media. Additions of 20-mM magnesium did not alter the disruptive effects of 1-mM calcium.

Calcium-induced alterations of neurofilaments also occurred in long, intact segments of rat sciatic nerve that had been excised, incubated in oxygenated nutrient media, and examined by electron

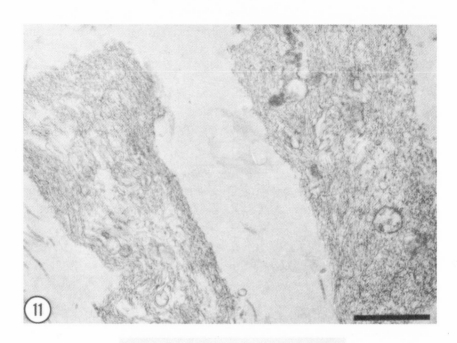

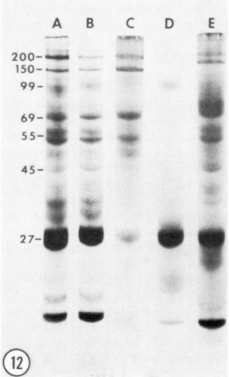

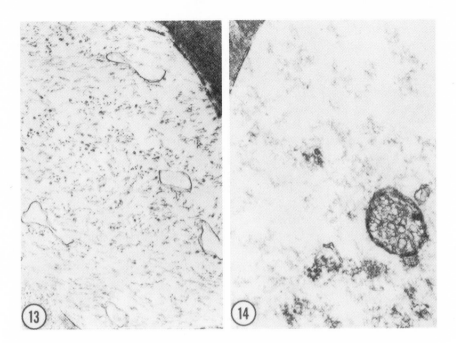

Figure 3-13. Transverse section of large myelinated fiber in rat sciatic nerve showing intact neurofilaments and microtubules after 3-day incubation in 0.1-M phosphate buffer, pH 7.0. (Magnified ×32,000.) (From reference 78. Used with permission.)

Figure 3-14. Transverse section of large myelinated fiber in rat sciatic nerve showing granular disintegration of neurofilaments and microtubules after 1-hour incubation in Ringer's solution. (Magnified ×32,000.) (From reference 78. Used with permission.)

microscopy or in whole-mount, teased fiber preparations (80). The calcium-dependent changes in these excised nerves were identical to those encountered in contralateral sciatic nerves that had been transected but left to degenerate *in vivo*. Degeneration of both transected and excised nerves occurred after the same latency period and were initiated by granular disintegration of neurofilaments and axonal microtubules. These changes did not occur in excised nerves that were incubated in media lacking calcium or in media in which calcium was

Figure 3-11. Representative section of purified axonal pellet from rat peripheral nerve showing portions of two axons in oblique orientation. Each axon is predominantly composed of punctate and linear profiles of neurofilaments. Neither microtubules nor axolemmae can be seen. Some collagen fibrils are seen along the lower left-hand margin. (Bar length 1 μm.) (From reference 64. Used with permission.)

Figure 3-12. SDS gel electrophoresis of proteins in rat peripheral nerve homogenate (A), rat nerve root homogenate (B), purified axonal pellet of rat nerve (C), myelin fraction from rat nerve (D), and human nerve root homogenate (E). Neurofilament triplet proteins of 68,000, 150,000, and 200,000 daltons are major constituents of neurofilament-rich axons. (From reference 64. Used with permission.)

chelated with EGTA. Furthermore, the same changes occurred at an accelerated rate when excised nerves were incubated in media containing elevated levels of calcium.

Characteristics of experimental calcium-induced axoplasmic degeneration similar to those seen in transected nerves have given rise to a calcium-influx hypothesis of Wallerian degeneration (Figure 3-15). According to this theory, the axonal cytoskeleton is normally maintained in an environment of very low calcium-ion concentration, similar to that measured in squid giant axons (3), thereby creating an enormous calcium gradient across the axolemma (Figure 3-15A). Focal disruption of the axolemma (e.g., transection or crush) leads to an immediate influx of calcium, causing local disintegration of neurofilaments and microtubules as depicted by Zelena and others (119) and schematically outlined in Figure 15B. This local influx of calcium and spread of axoplasmic degeneration remain limited, perhaps owing to the resealing of axolemma around collapsed axoplasm. The major influx of calcium into transected nerve fibers occurs across the axolemma (Figure 3-15C) after a latency period. Net entry of calcium into the axon raises axoplasmic calcium levels and triggers a widespread disintegration of neurofilaments and microtubules in transected nerve fibers.

Additional support for the calcium-influx hypothesis of Wallerian

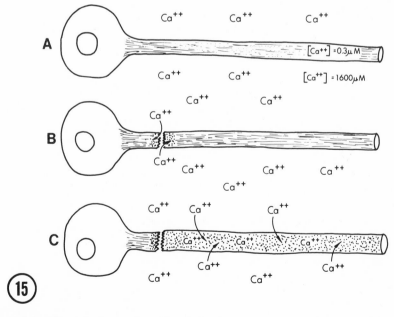

Figure 3-15. Schematic diagram for calcium-influx hypothesis of Wallerian degeneration. (From reference 80. Used with permission.)

degeneration is provided by studies of neurons *in vitro* (87). Transected neurites of dorsal-root-ganglia cultures show disintegration of their cytoskeleton only when calcium is present in the incubation media. Furthermore, these changes can be initiated by the addition of calcium to calcium-free media. Nevertheless, transient swellings of amputated fibers were not prevented by calcium depletion, indicating that the abnormal patterns of fluid and electrolyte exchange across axolemma of transected fibers (65) occur in the absence of calcium.

A correlate of the calcium-influx hypothesis of Wallerian degeneration is that an increased axolemmal permeability to calcium occurs during the initial latency period prior to morphological evidence of degeneration. It is also possible that release of calcium from internal axonal compartments may contribute to elevated axoplasmic calcium levels. The latter sources of calcium do not appear to be sufficient to generate widespread axonal breakdown when nerves are incubated in calcium-free media under energy-deprivational conditions (81). Furthermore, the acceleration of axonal breakdown during energy-depletional states indicates that the calcium-excluding properties of axolemma are energy dependent. Accordingly, anoxic conditions could be expected to hasten the process of Wallerian degeneration either locally or diffusely, possibly contributing to somatofugal or somatopetal progression of change (24,47). The preferential occurrence of Wallerian degeneration among fibers of different caliber could also result from tissue anoxia. For example, accelerated degeneration of small fibers occurs during nerve incubations under energy-deprivational conditions and is believed to reflect the greater surface area/ volume ratio in these fibers (81). Generally, however, high-energy metabolites appear to be maintained in transected nerves *in vivo* (34, 105).

Effects of calcium on invertebrate giant axons reveal close similarity to the interaction between calcium and mammalian nerve. For example, microinjections of calcium into squid giant axon produce a local opalescence and liquifaction of axoplasm (38), a change that also follows the local application of 1-mM calcium to extruded axoplasm (37). Studies of squid axons have shown that the axolemma serves as a very effective barrier to calcium ions, maintaining an intraaxoplasmic calcium-ion concentration of less than 0.3 μM (3). Small amounts of calcium that enter the squid axoplasm during depolarization (3) are quickly bound or removed and may be limited to the axoplasm immediately beneath the surface membrane (4). The calcium-excluding capacity of axonal membranes was also demonstrated in isolated squid axons incubated for 8 hours in media containing 10.7-mM calcium, during which time only 0.1 mM of [45]Ca entered the axon (39). Damaged regions of axonal membrane, however, were the sites of increased calcium permeability, which caused a visible cloudy alteration in the underlying axoplasm (29).

Calcium-Induced Alterations of Neurofilaments

Insight into the nature of neurofilaments and their disruption by calcium was gained by studying the latter phenomenon from both morphological and biochemical perspectives (94). These studies were conducted on peripheral nerve, and model systems in which the morphology of calcium-induced disruption of neurofilaments had been clearly established (80,81,86) were used. In all instances, the granular disintegration of neurofilaments by calcium was accompanied by the disappearance of neurofilament triplet proteins from tissue homogenates (94). These biochemical findings supported the newly emerged hypothesis of Hoffman and Lasek that mammalian neurofilaments were composed of three different proteins (40). Furthermore, the findings indicated that calcium caused profound alterations of neurofilament proteins, not just the disassembly of associated protein subunits. (For further discussion of calcium-dependent proteolysis of neurofilament proteins, see Chapter 5.)

The calcium-induced loss of neurofilament proteins occurs in excised peripheral nerves under the same conditions that lead to the granular disintegration of neurofilaments (93). Accordingly, loss of neurofilament proteins is accelerated by incubating nerves under energy-depletional states and by raising the calcium level of the incubation media. The same changes can be induced very quickly when excised nerves are subjected to membrane-disrupting procedures (e.g., freeze-thawing, addition of detergents) in the presence of calcium (Figure 3-16). Specificity of these changes for calcium is demonstrated by incubating nerve with the calcium ionophore A23187, which translocates calcium throughout the tissues. Addition of calcium to incubation media containing A23187 causes a selective loss of neurofilament triplet proteins and, to a lesser extent, of tubulin (Figure 3-16).

Loss of neurofilament triplet proteins provides a complementary as well as semiquantitative method of assessing calcium-induced disruption of neurofilaments. The demonstration of these protein alterations in transected nerves (93) reaffirms the similarity between changes of neurofilaments *in vivo* and the calcium-dependent alterations of neurofilaments that can be induced experimentally. Accordingly, the loss of neurofilament proteins coincides with the granular disintegration of neurofilaments in transected nerve. Biochemical changes occur during the same interval throughout the transected nerve, without evidence of proximodistal or distoproximal progression. At intermediate stages of neurofilament breakdown, losses of 68,000- and 150,000-mol wt neurofilament proteins are more advanced when compared with the loss of 200,000-mol wt neurofilament protein. By 72 hours, the three neurofilament proteins are no longer detectable in transected nerve (Figure 3-17). During the same time interval, most of the other proteins of peripheral nerve remain completely unaltered.

The nature of calcium-induced disruption of neurofilaments is most clearly revealed by studying the interaction of calcium with isolated neurofilaments. Both granular disintegration of neurofilaments and loss of neurofilament proteins can be reproduced by exposing crude preparations of neurofilaments to calcium (89). However, incubation

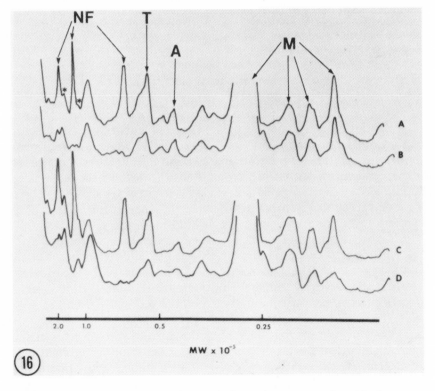

Figure 3-16. Densitometric tracings from SDS electrophoretograms of proteins in rat peripheral nerve segments incubated for 2 hours with ionophore, A23187, (top tracings, labeled A and B) or detergent, triton X-100 (bottom tracings, labeled C and D). Incubations were conducted in the presence (B and D tracings) or absence (A and C tracings) of calcium. Exposure to calcium caused selective loss of neurofilament triplet proteins (NF) and some loss of tubulin (T) but no alteration of actin (A) or myelin proteins (M). (From reference 94. Used with permission.)

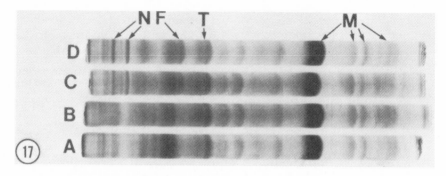

Figure 3-17. SDS electrophoretogram of nerve proteins in distal (A) and most proximal (B) segments of 72-hour transected rat sciatic nerve compared with proteins in the stump of intact sciatic nerve (C) and unsectioned axillary nerve (D). Neurofilament (NF) protein bands have disappeared from transected nerve, but tubulin (T) and myelin (M) proteins have remained largely unaltered. (From reference 93. Reproduced from *The Journal of Cell Biology*, 1978, vol. 78, pp. 369-378, by copyright permission of The Rockefeller University Press.)

of saline-washed neurofilaments with calcium fails to alter neurofilament structure (Figure 3-18) or their protein composition; this indicates that calcium does not act directly with neurofilaments to cause their disruption. On the other hand, parallel incubations with calcium of unwashed neurofilaments from the same preparation causes structural disintegration of neurofilament (Figure 3-19) and complete loss of neurofilament proteins. The marked difference between washed and unwashed neurofilaments in their reactivities to calcium is due to a tissue factor that separates from neurofilaments during their sedimentation in saline and can be recovered from the saline wash. The addition of this tissue factor to washed neurofilaments restores their susceptibility to calcium-induced disruption. The tissue factor is precipitable by $(NH_4)_2 SO_4$, is inactivated by heat (i.e., 60°C for 15 minutes), is inhibited by 1-mM p-chlormercuribenzoate (PCMB), and is able to disrupt neurofilaments between pH 6.0 and 8.0 and at 4° to 37°C (89).

The tissue factor that mediates the breakdown of mammalian neurofilaments has been interpreted as a calcium-activated neurofilament protease (89) and closely resembles a similar tissue factor that causes calcium-dependent disruption of neurofilaments from the giant invertebrate axons of squid (69) and *Myxiocola* (33,56). Neurofilament protease from rat tissue degrades radioiodinated neurofilament proteins at rates that are dependent upon enzyme and substrate concentrations and follow simple Michaelis-Menton kinetics (95). The enzyme can be activated by low concentrations of calcium (e.g., 10 μM), is inhibited by thiol reagents, and degrades casein and tubulin

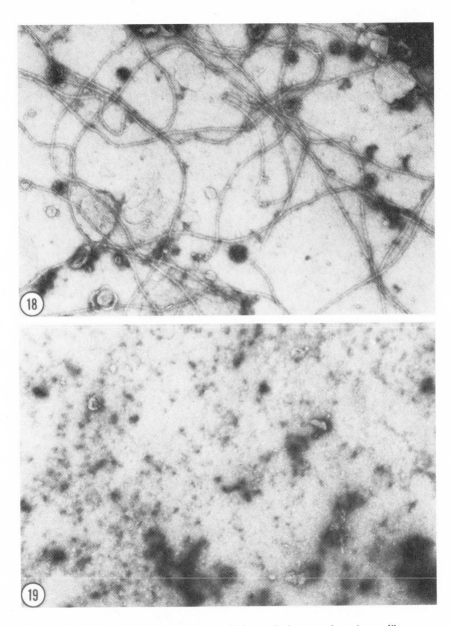

Figure 3-18. Saline-washed neurofilaments, which remain intact and can be readily demonstrated by negative stain after 2-hour incubation with 4-mM calcium. (From reference 89. Used with permission.)

Figure 3-19. Identical incubation of unwashed neurofilaments with calcium causing neurofilament breakdown so that only granular debris can be seen by negative stain. Neurofilament breakdown can be prevented by addition of 1-mM PCMB. (From reference 89. Used with permission.)

as well as neurofilament triplet proteins. Enzymatic inactivation by calcium may be due to autoproteolysis.

Studies of transverse frozen sections of rat sciatic nerve provide morphological evidence that the breakdown of neurofilaments in peripheral nerve is mediated by a calcium-activated, PCMB-sensitive axoplasmic protease (91). Neurofilaments within these frozen sections remain intact when incubated in calcium-free media prior to fixation and processing for ultrastructural examination. Simultaneous incubations of adjacent frozen sections with calcium results in widespread granular disintegration of neurofilaments, with variable collapse of myelinated nerve fibers around altered axoplasm. This calcium-mediated disruption of neurofilaments can be inhibited by brief (i.e., 5-second) preincubations in 1-mM PCMB prior to incubation with calcium. Furthermore, PCMB-inhibition can be partially reversed by a second preincubation with a reducing agent, dithioerythritol. The widespread loss of neurofilaments from frozen sections of nerve after incubation with calcium indicates that neurofilament protease is located throughout the axoplasm of peripheral nerve.

Peripheral nerve has also been used to characterize the properties of calcium-activated neurofilament protease *in situ*, with freeze-thawing procedures used to induce a calcium influx and to initiate enzymatic activity in segments of rat sciatic nerve (96). Enzymatic activity can be monitored by comparing the loss of neuorfilament triplet proteins from these nerve segments under different incubational conditions. The properties of calcium-activated neurofilament protease are outlined in Table 3-2. These properties closely resemble those of calcium-activated protease in other tissues, as indicated in Table 3-3.

REFERENCES

1. Anderton, B. H., M. Ayers, and R. Thorpe. 1978. Neurofilaments from mammalian central and peripheral nerve share certain polypeptides. *FEBS Lett.* 96:159-163.

2. Andres, K. H. 1961. Untersuchungen über den Feinbau von Spinalganglien. *Z. f. Zellforsch.* 55:1-48.

3. Baker, P. F., A. L. Hodgkin, and E. B. Ridgway. 1971. Depolarization and calcium entry in squid giant axons. *J. Physiol.* 218:709-755.

4. Baker, P. F., and W. W. Schlaepfer. 1978. Uptake and binding of calcium by axoplasm isolated from giant axons of *Loligo* and *Myxicola*. *J. Physiol.* 276:103-125.

5. Belocopitow, E., M. M. Appleman, and H. N. Torres. 1965. Factors affecting the activity of muscle glycogen synthetase: II. The regulation by Ca^{++}. *J. Biol. Chem.* 240: 3473-3478.

6. Berthold, C.-H. 1978. Morphology of normal peripheral axons. *In*: Physiology and Pathology of Axons. S. G. Waxman, editor. Raven Press, New York, pp. 3-63.

7. Black, M. M., and R. J. Lasek. 1979. Axonal transport of actin: Slow component b is the principal source of actin for the axon. *Brain Res.* 171:401-413.

Table 3-2
Properties of Neurofilament Protease Activity in Rat Peripheral Nerve

Substrates	Activity	Metal Inhibitors	Activity
NF protein (68,000 dalton MW)	+++	Mercury (1 mM)	0
NF protein (150,000 dalton MW)	+++	Zinc (1 mM)	0
NF protein (200,000 dalton MW)	++	Cadmium (1 mM)	0
tubulin	+	Copper (1 mM)	0
		Cobalt (1 mM)	0
Activators	**Activity**	Lead (1 mM)	++
		Silver (1 mM)	++
Calcium (0.05 to 10 mM)	+++	Aluminum (1 mM)	+++
Strontium (1 mM)	+++	Manganese (1 mM)	+++
Barium (1 mM)	++		
Lanthanum (1 mM)	+	**Organic Inhibitors**	**Activity**
pH	**Activity**	PCMB (1 mM)	0
		Iodoacetate (1 mM)	0
6.0	+	N-ethyl maleimide (1 mM)	0
6.5	++	TLCK (1 mM)	0
7.0	+++	TPCK (1 mM)	0
7.5	+++	PMSF (1 mM)	++
8.0	+++	Soybean trypsin inhibitor	+++
8.5	++		
9.0	+		

Source: Reference 96. Used with permission.

Note: Proteolytic activity was based upon comparative losses of neurofilament proteins from electrophoretic profiles of whole nerve homogenates when nerves were incubated under different incubational conditions. Maximum proteolytic activity (+++) corresponded to complete loss of protein bands, whereas moderate (++), mild (+), or no (0) enzymatic activity was assessed by moderate, mild, or no loss of neurofilament proteins from gels when subjected to densitometric scan.

8. Bodian, D. 1968. Development of fine structure of spinal cord in monkey fetuses: II. Pre-reflex period to period of long intersegmental reflexes. J. Comp. Neurol. 133:113-166.

9. Brown, B. A., R. A. Nixon, P. Strocchi, and C. A. Marotta. 1981. Characterization and comparison of neurofilament proteins from rat and mouse CNS. J. Neurochem. 36: 143-153.

10. Bunge, M. B. 1973. Fine structure of nerve fibers and growth cones of isolated sympathetic neurons in culture. J. Cell. Biol. 56:713-735.

11. Chiu, R.-C., B. Korey, and W. T. Norton. 1980. Intermediate filaments from bovine, rat, and human CNS: Mapping analysis of the major proteins. J. Neurochem. 34:1149-1159.

12. Czosnek, H., and D. Soifer, 1980. Comparison of the proteins of 10 nm filaments from rabbit sciatic nerve and spinal cord by electrophoresis in two dimensions. FEBS Lett. 117:175-178.

13. Czosnek, H., D. Soifer, K. Mack, and H. M. Wisniewski. 1981. Similarity of neurofilament proteins from different parts of the rabbit nervous system. Brain Res. 216:387-398.

14. Czosnek, H., D. Soifer, and H. M. Wisniewski. 1980. Heterogeneity of intermediate filament proteins from rabbit spinal cord. Neurochem. Res. 5:777-793.

15. Czosnek, H., D. Soifer, and H. M. Wisniewski. 1980. Studies on the biosynthesis of neurofilament proteins. J. Cell Biol. 85:726-734.

16. Dahl, D. 1979. The cyanogen bromide peptide maps of neurofilament polypeptides

Table 3-3

Common Properties of Calcium-Activated Proteases

Source	Substrate	Activation by 0.1-1.0 mM Ca	Inactivation by Ca	Activation by Sr, Mn, or Ba	pH Range	Inhibitors		
						Sulfhydral Reagents	Serine Reagents	Heavy Metals
Nerve	Neurofilaments	+		+	6-9	+	–	+
Brain	Casein (35)	+	+	+	6-8	+	–	+
	Histone phosphorylase (43)	+	+	+	7-8	+		
	Phosphorylase b (25)	+	+					
Muscle	Casein (21,45,75)	+	+	+	6-8	+	–	+
	Myofibrillary proteins (21,95)	+	+		6-8	+	–	+
	Filamin (17)	+						
	Phosphorylase b (25,42,63)	+	+	+	6-9	+	–	+
	Glycogen synthetase (5)	+			6-9			
Smooth Muscle	Estrogen receptor protein (73)	(4 mM)	+	+	6-10	+	–	
	Casein (73)	(4 mM)	+	+	6-10	+	–	
Platelets	Azocasein (71)	(2 mM)			6	+	–	

Source: Reference 96. Used with permission.

in axonal preparations isolated from bovine brain are different. *FEBS Lett. 103*:144-147.

17. Davies, P. J. A., D. Wallach, M. C. Willingham, and I. Pastan. 1978. Filamin-actin interaction: Dissociation of binding from gelatin by Ca²⁺-activated proteolysis. *J. Biol. Chem. 253*:4036-4042.

18. Davison, P. F., and B.-S. Hong. 1977. Structural homologies in mammalian neurofilament proteins. *Brain Res. 134*:287-295.

19. Davison, P. F., and B. Winslow. 1974. The protein subunit of calf brain neurofilament. *J. Neurobiol. 5*:119-133.

20. Day, W. A., and D. S. Gilbert, 1972. X-ray diffraction pattern of axoplasm. *Biochim. Biophys. Acta 285*:503-506.

21. Dayton, W. R., W. J. Reville, D. E. Goll, and M. H. Stromer. 1976. A Ca²⁺-activated protease possibly involved in myofibrillary protein turnover: Partial characterization of the purified enzyme. *Biochemistry 15*:2159-2167.

22. DeVries, G. H., L. F. Eng, D. L. Lewis, and M. G. Hadfield. 1976. The protein composition of bovine myelin-free axons. *Biochim. Biophys. Acta 439*:133-145.

23. DeVries, G. H., W. T. Norton, and C. S. Raine, 1972. Axons: Isolation from mammalian central nervous system. *Science 175*:1370-1372.

24. Donat, J. R., and H. M. Wisniewski. 1973. The spatio-temporal pattern of Wallerian degeneration in mammalian peripheral nerves. *Brain Res. 53*:41-53.

25. Drummond, G. I., and L. Duncan. 1968. On the mechanism of activation of phosphorylase b kinase by calcium. *J. Biol. Chem. 243*:5532-5538.

26. Elfvin, L.-G. 1961. The ultrastructure of the nodes of Ranvier in cat sympathetic nerve fibers. *J. Ultrastruct. Res. 5*:374-387.

27. Ellisman, M. H., and K. R. Porter. 1980. Microtrabecular structure of the axoplasmic matrix: Visualization of cross-linking structures and their distribution. *J. Cell. Biol. 87*:464-479.

28. Filliatreau, G., L. DiGiamberardino, A. Delacourte, F. Boutteau, and G. Biserte. 1981. Composition polypeptidique des neurofilaments d'oiseaux. *Biochimie 63*:369-371.

29. Fluckiger, E., and R. D. Keynes. 1955. The calcium permeability of *Loligo* axons. *J. Physiol. 128*:41-42p.

30. Friede, R. L., and T. Samorajski. 1970. Axon caliber related to neurofilaments and microtubules in sciatic nerve fibers of rats and mice. *Anat. Rec. 167*:379-388.

31. Gilbert, D. S. 1972. Helical structure of *Myxicola* axoplasm. *Nature New Biol. 237*: 195-224.

32. Gilbert, D. S. 1975. Axoplasm architecture and physical properties as seen in the *Myxicola* giant axon. *J. Physiol. 253*:257-301.

33. Gilbert, D. S., B. J. Newby, and B. H. Anderton. 1975. Neurofilament disguise, destruction, and discipline. *Nature 265*:586-589.

34. Greengard, P., F. Brink, and P. S. Colwick. 1954. Some relationships between action potential, oxygen consumption, and coenzyme content in degenerating peripheral axons. *J. Cell Comp. Physiol. 44*:395-420.

35. Guroff, G. 1964. A neutral, calcium-activated proteinase from the soluble fraction of rat brain. *J. Biol. Chem. 239*:149-155.

36. Ha, H. 1970. Axonal bifurcation in the dorsal root ganglion of the cat: A light and electron microscopic study. *J. Comp. Neurol. 140*:227-240.

37. Hodgkin, A. L., and B. Katz. 1949. The effect of calcium on the axoplasm of giant nerve fibers. *J. Exp. Biol. 26*:292-294.

38. Hodgkin, A. L., and R. D. Keynes. 1956. Experiments on the injection of substances into squid giant axons by means of microsyringe. *J. Physiol. 131*:592-616.

39. Hodgkin, A. L., and R. D. Keynes. 1957. Movements of labelled calcium in squid giant axons. *J. Physiol. 138*:253-281.

40. Hoffman, P. N. and R. J. Lasek. 1975. The slow component of axonal transport: Identification of major structural polypeptides of the axon and their generality among mammalian neurons. *J. Cell Biol. 66*:351-366.

41. Huneeus, F. C., and P. F. Davison. 1970. Fibrillar proteins from squid axons: I. Neurofilament protein. *J. Mol. Biol.* 52:415-428.

42. Huston, R. B., and E. G. Krebs. 1968. Activation of skeletal muscle phosphorylase kinase by Ca^{2+}: II. Identification of the kinase activating factor as a proteolytic enzyme. *Biochemistry* 7:2116-2122.

43. Inoue, M., A. Kishimoto, Y. Takai, and Y. Nishizuka. 1977. Studies on a cyclic nucleotide-independent protein kinase and its proenzyme in mammalian tissues: II. Proenzyme and its activation by calcium-dependent protease from rat brain. *J. Biol. Chem.* 252:7610-7616.

44. Iqbal, K., I. Grundke-Iqbal, H. M. Wisniewski, and R. D. Terry. 1977. On the neurofilament and neurotubule proteins from human autopsy tissue. *J. Neurochem.* 29:417-424.

45. Ishiura, S., H. Murofushi, K. Susuki, and K. Imahori. 1978. Studies of calcium-activated neutral protease from chicken skeletal muscle: I. Purification and characterization. *J. Biochem.* 84:225-230.

46. Jokusch, H., B. M. Jokusch, and M. M. Burger. 1979. Nerve fibers and their interactions with non-neural cells visualized in immunofluorescence. *J. Cell Biol.* 80:629-641.

47. Joseph, B. S. 1973. Somatofugal events in Wallerian degeneration: A conceptual overview. *Brain Res.* 59:1-18.

48. Kadota, T., and K. Kadota. 1972. A simple and fundamental quadrate structure of neurofilament in various nerve endings and its relation to intervesicular networks and dense projections. *J. Elec. Micros.* 21:218.

49. Kelly, P. T., and M. W. Luttges. 1975. Electrophoretic separation of nervous system proteins on exponential gradient polyacrylamide gels. *J. Neurochem.* 23:1077-1079.

50. Kidd, M. 1963. Paired helical filaments in electron microscopy of Alzheimer's disease. *Nature* 197:192-193.

51. Komiya, H., and M. Kurokawa. 1978. Asymmetry of protein transport in two branches of bifurcating axons. *Brain Res.* 139:354-358.

52. Krishnan, N., I. R. Kaiserman-Abramof, and R. J. Lasek. 1979. Helical substructure of neurofilaments from *Myxicola* and squid giant axons. *J. Cell Biol.* 82:323-335.

53. Kurczmarski, E. R., and J. L. Rosenbaum. 1979. Studies of the organization and localization of actin and myosin in neurones. *J. Cell Biol.* 80:365-371.

54. Lasek, R. J. 1981. The dynamic ordering of neuronal cytoskeleton. *Neurosciences Res. Prog. Bull.* 19:7-31.

55. Lasek, R. J., and P. N. Hoffman. 1976. The neuronal cytoskeleton, axonal transport, and axonal growth. *In*: Cell Motility. Vol. 3, Microtubules and Related Proteins. R. Goldman, T. Pollard, and J. Rosenbaum, editors. Cold Spring Harbor Laboratory, Cold Spring Harbor, N.Y., pp. 1021-1049.

56. Lasek, R. J., N. Krishnan, and I. R. Kaiserman-Abramof. 1979. Identification of the subunit proteins of 10-nm neurofilaments isolated from axoplasm of squid and *Myxicola* giant axons. *J. Cell Biol.* 82:336-346.

57. LeBeux, Y. J. 1973. An ultrastructural study of the synaptic densities, nematosomes, neurotubules, neurofilaments, and of a further three-dimensional filamentous network as disclosed by the E-PTA staining procedure. *Z. Zellforsch. Mikrosk. Anat.* 143:239-272.

58. LeBeux, Y. J., and J. Willemot. 1975. An ultrastructural study of the microfilaments in rat brain by means of E-PTA staining and heavy meromyosin labeling: II. The synapses. *Cell Tissue Res.* 160:37-68.

59. Liem, R. K. H., S.-H. Yen, G. D. Salomon, and M. L. Shelanski. 1978. Intermediate filaments in nervous tissues. *J. Cell Biol.* 79:637-645.

60. Lyser, K. M. 1968. Early differentiation of motor neuroblasts in the chick embryo as studied by electron microscopy: II. Microtubules and neurofilaments. *Devel. Biol.* 17:117-142.

61. Metuzals, J. 1969. Configuration of a filamentous network in the axoplasm of the squid (*Loligo pealii* L.) giant nerve fiber. *J. Cell Biol.* 43:480-505.

62. Metuzals, J., and W. E. Mushynski. 1974. Electron microscope and experimental investigations of the neurofilamentous network in Deiters' neurons: Relationship with the cell surface and nuclear pores. *J. Cell Biol.* 61:701-702.

63. Meyer, W. W., E. H. Fischer, and E. G. Krebs. 1964. Activation of skeletal muscle phosphorylase b kinase by Ca^{2+}. *Biochemistry* 3:1033-1039.

64. Micko, S., and W. W. Schlaepfer. 1978. Protein composition of axons and myelin from rat and human peripheral nerves. *J. Neurochem.* 30:1041-1049.

65. Mire, J. J., W. J. Hendelman, and R. P. Bunge. 1970. Observations on a transient phase of focal swelling in degenerating unmyelinated nerve fibers. *J. Cell Biol.* 45:9-22.

66. Mori, H., Y. Komiya, and M. Kurokawa. 1979. Slowly migrating axonal polypeptides: Inequalities in their rate and amount of transport between two branches of bifurcating axons. *J. Cell Biol.* 82:174-184.

67. Mori, H., and M. Kurokawa. 1979. Purification of neurofilaments and their interaction with vinblastine sulphate. *Cell Struct. Funct.* 4:163-167.

68. Palay, S. L., C. Sotelo, A. Peters, and P. M. Orkland. 1968. The axon hillock and the initial segment. *J. Cell Biol.* 38:193-201.

69. Pant, H. C., S. Terakawa, and H. Gainer. 1979. A calcium-activated protease in squid axoplasm. *J. Neurochem.* 32:99-102.

70. Peters, A., and J. E. Vaughn. 1967. Microtubules and filaments in the axons and astrocytes of early postnatal rat optic nerves. *J. Cell Biol.* 32:113-119.

71. Phillips, D. R., and M. Jakabova. 1977. Ca^{2+}-dependent protease in human platelets: Specific cleavage of platelet polypeptides in the presence of added Ca^{2+}. *J. Biol. Chem.* 252: 5602-5605.

72. Plancke, Y., A. Delacourte, and G. Biserte. 1981. Preparation of neurofilaments from three different sources. *Biochimie* 63:365-367.

73. Puca, G. A., E. Nola, V. Sica, and E. Bresciani. 1977. Estrogen binding proteins of calf uterus: Molecular and functional characterization of the receptor transforming factor: A Ca^{2+}-activated protease. *J. Biol. Chem.* 252:1358-1366.

74. Raju, T., A. Bignami, and D. Dahl. 1981. *In vivo* and *in vitro* differentiation of neurons and astrocytes in the rat embryo. *Devel. Biol.* 85:344-357.

75. Reddy, M. K., J. D. Etlinger, M. Rabinowitz, D. A. Fischman, and R. Zak. 1975. Removal of Z-lines and a-actinin from isolated myofibrils by a calcium-activated neural protease. *J. Biol. Chem.* 250:4278-4284.

76. Robertson, J. D. 1959. Preliminary observations on the ultrastructure of nodes of Ranvier. *J. Biophys. Biochem. Cytol.* 4:349-364.

77. Runge, M. S., W. W. Schlaepfer, and R. C. Williams, Jr. 1981. Isolation and characterization of neurofilaments from mammalian brain. *Biochemistry* 20:170-175.

78. Schlaepfer, W. W. 1971. Experimental alteration of neurofilaments and neurotubules by calcium and other ions. *Exp. Cell Res.* 67:73-80.

79. Schlaepfer, W. W. 1971. Stabilization of neurofilaments by vincristine sulfate in low ionic strength media. *J. Ultrastruct. Res.* 36:367-374.

80. Schlaepfer, W. W. 1974. Calcium-induced degeneration of axoplasm in isolated segments of rat peripheral nerve. *Brain Res.* 69:203-215.

81. Schlaepfer, W. W. 1974. Effects of energy deprivation on Wallerian degeneration in isolated segments of rat peripheral nerve. *Brain Res.* 78:71-81.

82. Schlaepfer, W. W. 1977. Immunological and ultrastructural studies of neurofilaments isolated from rat peripheral nerve. *J. Cell Biol.* 74:226-240.

83. Schlaepfer, W. W. 1977. Studies on the isolation and substructure of mammalian neurofilaments. *J. Ultrastruct. Res.* 61:149-157.

84. Schlaepfer, W. W. 1978. Deformation of isolated neurofilaments and the pathogenesis of neurofibrillary pathology. *J. Neuropathol. Exp. Neurol.* 38:244-254.

85. Schlaepfer, W. W. 1978. Observations on the disassembly of isolated mammalian neurofilaments. *J. Cell Biol.* 76:50-56.

86. Schlaepfer, W. W. 1978. Structural alterations of peripheral nerve induced by the calcium ionophore, A23187. *Brain Res. 136*:1-9.

87. Schlaepfer, W. W., and R. P. Bunge. 1973. The effects of calcium ion concentration on the degeneration of amputated axons in tissue culture. *J. Cell Biol. 59*:456-470.

88. Schlaepfer, W. W., and L. A. Freeman. 1978. Neurofilament proteins of rat peripheral nerve and spinal cord. *J. Cell Biol. 78*:653-662.

89. Schlaepfer, W. W., and L. A. Freeman. 1980. Calcium-dependent degradation of mammalian neurofilaments by soluble tissue factor(s) from rat spinal cord. *Neuroscience 5*: 2305-2314.

90. Schlaepfer, W. W., L. A. Freeman, and L. F. Eng. 1979. Studies of human and bovine spinal nerve roots and the outgrowth of CNS tissue into the nerve root entry zone. *Brain Res. 177*:219-229.

91. Schlaepfer, W. W., and M. B. Hasler. 1979. Characterization of the calcium-induced disruption of neurofilaments in rat peripheral nerve. *Brain Res. 168*:299-309.

92. Schlaepfer, W. W., and R. G. Lynch. 1977. Immunofluorescence studies of neurofilaments in the rat and human peripheral and central nervous system. *J. Cell Biol. 74*:241-250.

93. Schlaepfer, W. W., and S. Micko. 1978. Chemical and structural changes of neurofilaments in transected rat sciatic nerve. *J. Cell Biol. 78*:369-378.

94. Schlaepfer, W. W., and S. Micko. 1979. Calcium-dependent alterations of neurofilament proteins of rat peripheral nerve. *J. Neurochem. 32*:211-219.

95. Schlaepfer, W. W., and U.-J. Zimmerman. 1981. The breakdown of neurofilaments by calcium-activated proteolysis. *J. Neuropathol. Exp. Neurol. 40*:315.

96. Schlaepfer, W. W., U.-J. Zimmerman, and S. Micko. 1981. Neurofilament proteolysis in rat peripheral nerve: Homologies with calcium-activated proteolysis of other tissues. *Cell Calcium 2*:235-250.

97. Schook, W. J., and W. T. Norton. 1975. On the composition of axonal neurofilaments. *Trans. Am. Soc. Neurochem. 6*:214 (abs.).

98. Schook, W. J., and W. T. Norton. 1976. Neurofilaments account for the lipid in myelin-free axons. *Brain Res. 118*:517-522.

99. Shaw, G., M. Osborn, and K. Weber. 1981. Arrangement of neurofilaments, microtubules, and microfilament-associated proteins in cultured dorsal root ganglia cells. *Eur. J. Cell Biol. 24*:20-27.

100. Schecket, G., and R. J. Lasek. 1981. Preparation of neurofilament protein from guinea pig peripheral nerve and spinal cord. *J. Neurochem. 35*:1335-1344.

101. Shelanski, M. L., S. Albert, G. H. DeVries, and W. T. Norton. 1971. Isolation of filaments from brain. *Science 174*:1242-1245.

102. Smith, D. E. 1973. The location of neurofilaments and microtubules during the postnatal development of Clarke's nucleus in the kitten. *Brain Res. 55*:41-53.

103. Smith, R. S. 1973. Microtubules and neurofilament densities in amphibian spinal nerve fibers: Relationship to axoplasmic transport. *Can. J. Physiol. Pharmacol. 51*:798-806.

104. Steinert, P. M., W. W. Idler, and R. D. Goldman. 1980. Intermediate filaments of baby hamster kidney (BHK-21) cells and bovine epidermal keratinocytes have similar ultrastructures and subunit domain structures. *Proc. Natl. Acad. Sci. USA 77*:4534-4538.

105. Steward, M. A., J. V. Passonneau, and O. H. Lowry. 1965. Substrate changes in peripheral nerve during ischaemia and Wallerian degeneration. *J. Neurochem. 12*:719-727.

106. Tapscott, S. J., G. S. Bennett, Y. Toyama, F. Kleinbart, and H. Holtzer. 1981. Intermediate filament proteins in the developing chick spinal cord. *Devel. Biol. 86*:40-54.

107. Tennyson, V. M. 1970. The fine structure of the axon and growth cone of the dorsal root neuroblast of the rabbit embryo. *J. Cell Biol. 44*:62-79.

108. Tennyson, V. M. 1975. Light and electron microscopy of dorsal root and sympathetic ganglia. *In*: Peripheral Neuropathy. P. J. Dyck, P. K. Thomas, and E. H. Lambert, editors. W. B. Saunders, Philadelphia, pp. 74-103.

109. Thorpe, R., A. Delacourte, and B. H. Anderton. 1979. The isolation of brain 10 nm filament polypeptides from urea-extract of brain white matter. *FEBS Lett. 103*:148-151.

110. Thorpe, R., A. Delacourte, M. Ayers, C. Bullock, and B. H. Anderton. 1979. The polypeptides of isolated brain 10 nm filaments and their association with polymerized tubulin. *Biochem. J. 181*:275-284.

111. Willard, M. 1977. The identification of two intra-axonally transported polypeptides resembling myosin in some respects in the rabbit visual system. *J. Cell Biol. 75*:1-11.

112. Willard, M., M. Wiseman, J. Levine, and P. Skene. 1979. Axonal transport of actin in rabbit retinal ganglion cells. *J. Cell Biol. 81*:581-591.

113. Wisniewski, H. M., H. K. Narang, and R. D. Terry. 1976. Neurofibrillary tangles of paired helical filaments. *J. Neurol. Sci. 27*:173-181.

114. Wolosewick, J. J., and K. R. Porter. 1979. Microtrabecular lattice of the cytoplasmic ground substance: Artifact or reality? *J. Cell Biol. 82*:114-139.

115. Wuerker, R. B. 1970. Neurofilaments and glial filaments. *Tiss. Cell 2*:1-9.

116. Wuerker, R. B., and J. B. Kirkpatrick. 1972. Neuronal microtubules, neurofilaments, and microfilaments. *Int. Rev. Cytol. 33*:45-75.

117. Wuerker, R. B., and S. L. Palay. 1969. Neurofilaments and microtubules in the anterior horn cells of the rat. *Tiss. Cell 1*:387-402.

118. Yen, S. H., D. Dahl, M. Schachner, and M. L. Shelanski. 1976. Biochemistry of the filaments of the brain. *Proc. Natl. Acad. Sci. USA 73*:529-533.

119. Zelena, J., L. Lubinska, and E. Guttmann. 1968. Accumulation of organelles at the ends of interrupted axons. *Z. Zellforsch. 91*:200-219.

120. Zenker, W., and E. Högl. 1976. The prebifurcation section of the axon of the rat spinal ganglion cell. *Cell Tissue Res. 165*:345-363.

121. Zenker, W., R. Mayr, and H. Gruber. 1974. Neurotubules: Different densities in peripheral motor and sensory nerve fibres. *Experientia 31*:318-320.

Chapter 4

Neurofilaments
and Axonal Transport

Mark Willard

The neurofilamentous cytoskeleton is a major architectural element of many types of neurons. Understanding its repertoire of behavior is important from two points of view. On the one hand, the neurofilaments are uniquely neuronal structures whose constituent proteins are not found in other types of cells; this specificity suggests that certain aspects of their function reflect the solutions to problems that are peculiar to neurons. On the other hand, the neurofilaments are an example of a more general class of morphologically similar filaments, designated intermediate filaments, found in many types of cells. As such, their behavior may reveal principles of the organization and dynamics of cytoskeletons that will also apply to non-neuronal intermediate filaments.

An unusual feature of neurons that recommends them as a model system for studying the behavior of the cytoskeleton is their extreme asymmetry; the axon, which may contain most of the cytoplasm and cytoskeleton (more than 99% in neurons with long axons), can project as far as several meters from the cell body. Because the process of protein synthesis is restricted to the cell body, the axon is absolutely dependent upon materials, including cytoskeletal proteins, that are elaborated in the cell body and conveyed down the axon by the process of axonal transport. In most cells, the distance over which

This chapter benefited from the generous contribution of time and perceptive comments from C. Baitinger, R. Cheney, M. Glicksman, Dr. K. Meiri, and C. Simon and the typing of Jan Hoffmann. I am grateful to Dr. Nobutaka Hirokawa for the use of his micrographs in Figure 4-1. This work was supported by grant EY02682 and a Jerry Lewis Neuromuscular Research Center grant.

the intermediate filamentous cytoskeleton must function between its site of synthesis and site of final disassembly is microscopic; in many neurons, this distance is greatly expanded, providing a macroscopic separation of synthesis in the cell body, function in the axon, and final disassembly in the terminal regions of the axon. Because of this spatial separation of the sequential stages in the life cycle of neurofilaments, neurons provide a uniquely appropriate system in which to consider the following questions: How is an intermediate filamentous cytoskeleton supplied to the cytoplasm of a mature cell? How does it function between the time of its inception and the time of its demise? How is it eventually disassembled? Are special mechanisms required to elaborate the filamentous cytoskeleton during cell development? The answers to these questions are not yet known in molecular detail. It is the purpose of this chapter to consider certain phenomena involved in the axonal transport of neurofilaments and to suggest how they may be related to the answers for these questions.

Axoplasm Structure and Axonal Transport

Electron microscopy has provided a static representation of the structure of axoplasm. According to this representation, neurofilaments are ubiquitous organelles with diameters of approximately 100 Å that run parallel to the long axis of the axon for distances of many micrometers. The neurofilaments are extensively cross-linked to each other and to microtubules, mitochondria, endoplasmic reticulum, and vesicles by thin cross-bridges with diameters of about 20 Å (18,34,55,79,86,89) (Figure 4-1). The linear elements and the interconnecting cross-bridges form a complex lattice that has been subject to different interpretations, depending upon the frame of reference suggested by the particular method of analysis. According to the microtrabecular interpretation, derived initially from high-voltage electron micrographs of cells (18,86), the frame of reference is the interlinking cross-bridges, or trabeculae, which are considered to compose an autonomous organelle; the straight filaments and tubules serve to form a supporting scaffolding. According to an alternative interpretation (34,55,79,89), derived in part from observations of platinum replicas of rapidly frozen, fractured, and etched specimens, the frame of reference is the filaments and tubules; the cross-bridges serve to link other constituents of the axon to them. Although they differ in their emphasis upon the interlinking material, or the straight filaments and tubules, and in their speculations as to the nature of the interlinking materials, these two views concur that the axonal cytoskeleton comprises a complex lattice of linear elements and crossbridges, which enwebs the other organelles of the axon (Figure 4-1).

A

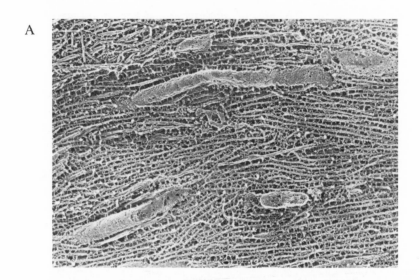

B

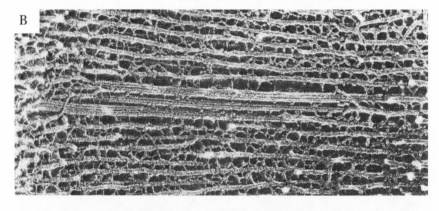

C

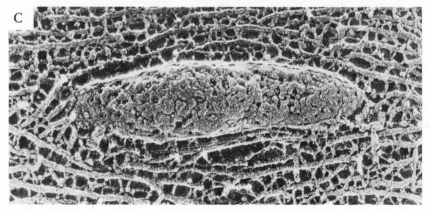

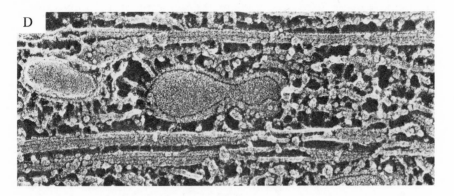

Figure 4-1. The neurofilamentous cytoskeleton as represented by electron microscopy of platinum replicas of segments of frog sciatic nerve that had been rapidly frozen and freeze-etched (34): (A) mitochondria embedded in a matrix of neurofilaments that are cross-linked to each other, to microtubules, and to the mitochondria; (B) the extensive cross-bridging of neurofilaments and microtubules (larger-diameter organelles in the central region of the figure) that give rise to the NF-MT lattice; (C) the cross-bridges between neurofilaments and a mitochondrion; (D) the links between membrane-bound organelles and microtubules. (These electron micrographs were provided by Dr. Nobutaka Hirokawa [34]. Reproduced from *The Journal of Cell Biology*, 1982, vol. 94, pp. 129-142, by copyright permission of The Rockefeller University Press.)

The problem of how the neuronal cytoskeleton is renewed and maintained is intriguing, considering that newly synthesized elements are available only at one end of the axon column, which, in the case of the peripheral nerves of large mammals, may be more than a meter in length. The extensive cross-linking of the filaments raises the additional problem of how other transported materials find their way through the intricate maze created by the filaments and their cross-bridges. A particularly intriguing question is whether the cross-bridges between membrane-bounded organelles and the linear filaments and tubules are static structures that serve to hold the organelles in place or dynamic structures that have been frozen in the act of moving these organelles down the axon. These questions cannot be answered solely by the static images produced by electron microscopy.

The dynamic nature of axoplasm can be appreciated by direct microscopic observation of living axons, which has revealed the rapid movements (several micrometers per second) of unidentified particles and the slower movements of mitochondria (12,20). In addition, the movements of individual molecules down axons can be observed indirectly by labeling the cell body with radioactive precursors of axonal materials and subsequently recovering the radioactive product from a segment of the axon. The time required by the newly synthesized materials to reach a particular segment or to pass between two segments can be used to calculate the velocity of their movement within the axon. The analysis of radioactive, axonally transported polypeptides by electrophoretic techniques has indicated that axonally transported polypeptides can be divided into at least five groups according to their maximum velocities (1). Group I (velocity greater than 240 mm per day) includes such proteins as the Na^+-K^+-ATPase and acetylcholinesterase, as well as phospholipids, glycolipids, and glycoproteins (31). These materials serve to maintain the neuronal plasma membrane. Protein I, a constituent of synaptic vesicles, also moves at the velocity of group I (1). Group II (velocity approximately 40 mm per day) comprises a heterogeneous collection of proteins including mitochondrial proteins. This group apparently is generated by the movement of mitochondria and other, as yet unidentified organelles down the axon (52). Two high-molecular-weight (250,000 and 240,000) group-II polypeptides, designated fodrin (from the Greek for lining), are most concentrated beneath the plasma membrane of the axon, as though they formed a lining of the entire neuronal plasma membrane (48). Fodrin appears to be transported in association with more slowly moving groups, as well as with group II.

The mechanism for conveying materials rapidly (i.e., at the velocities of groups I and II) is independent of the cell body (14,53,63) and requires the expenditure of energy, most likely in the form of ATP (17,22,41,61,62). The process is inhibited by agents that disrupt microtubules (20) and actin filaments (30,39), suggesting that the transport mechanism depends upon intact microtubules and microfilaments (80). A portion of the rapidly transported proteins may be conveyed down the axon in association with a vehicle that appears in electron micrographs as an elongated "vesiculo-tubular" structure (89). Although numerous mechanisms have been proposed to explain the rapid transport process, the details remain elusive (20,80).

The third and fourth groups of axonally transported polypeptides (velocities of 2 to 8 mm per day) comprise a large number of proteins, including actin (2,98), two forms of myosinlike proteins (93), glycolytic enzymes (4), calmodulin (5,19), clathrin (23), and additional fodrin. The nature of the slowly moving structures with which these proteins are associated has not been demonstrated definitively; it has been speculated that these groups are generated by the movement of an axoplasmic matrix, such as the cross-linking lattice or microtrabeculae (2,98). At least some of the proteins of these groups are most concentrated in the peripheral region of the axon cylinder underlying the plasma membrane (21). The inclusion in groups III and IV of actin and myosinlike proteins, whose counterparts in muscle serve to generate mechanical forces, suggests the possibility that this subaxolemmal component of groups III and IV may include the machinery used for mechanochemical purposes in the axon, such as the transport of other materials (98).

The composition of group V (velocity about 1 mm per day) is quite simple; it includes α and β tubulin (37), additional fodrin (47, 95), and the three polypeptides (195,000-, 145,000-, and 73,000-mol wt in the rabbit visual system) (95) associated with the 100-Å neurofilaments (37) (Figure 4-2). Henceforth, these neurofilament polypeptides will be referred to collectively as the triplet polypeptides (37) and individually as H, M, and L, indicating the high-, middle-, and low-molecular-weight filament polypeptides, respectively.

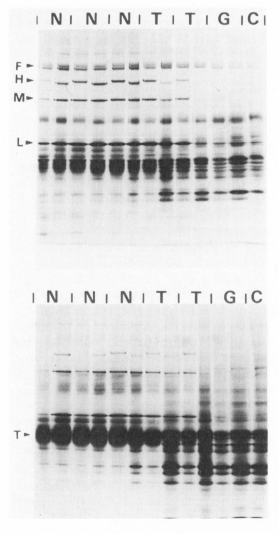

Figure 4-2. Group V polypeptides in segments of the optic nerve (N) and optic tract (T) sequentially more distal (left to right) from the eye, 35 days after the cell bodies of the retinal ganglion cells were labeled by an intravitreal injection of [^{35}S] methionine. The figure is an autoradiograph of a gel containing proteins from a particulate (upper) or soluble (lower) fraction of the tissue. F indicates fodrin; H, M, and L indicate the three neurofilament polypeptides, and T indicates tubulin. Each pair of samples includes the equivalent tissue from two different strains of rabbits: New Zealand white on the left of each pair and X/J on the right. The altered electrophoretic mobility of H in strain X/J reflects a genetically determined polymorphism of this polypeptide. G and C indicate the lacteral geniculate nucleus and superior colliculus, respectively, which contain the synaptic endings of retinal ganglion cell axons.

Association of the Group V Triplet with Neurofilaments

A relationship between the group V triplet polypeptides and neurofilaments was first suggested by Hoffman and Lasek (37,44). They reasoned that the labeled triplet proteins, which coelectrophorese with major polypeptides of the sciatic nerve, must be components of a major axonal structure, and they inferred that the triplet proteins are components of the 100-Å neurofilaments. Subsequent observations have provided strong support for this conclusion: during periods of experimentally induced neuronal degeneration, neurofilaments and triplet polypeptides accumulate and disappear in synchrony (78,81); partially purified preparations of intermediate filaments from peripheral and central nervous tissues are enriched in polypeptides that electrophoretically resemble the triplet (51,76); highly purified, dissociated, tripletlike polypeptides can form filamentous structures *in vitro* (24,50,56).

Although these observations leave little doubt that neurofilaments comprise three polypeptides that resemble the group V triplet polypeptides in electrophoretic mobility, they do not unequivocally equate the radiolabeled group V triplet with the neurofilament polypeptides. In the case of H, a genetic polymorphism has provided evidence that the transported polypeptide and the filament polypeptide are products of the same gene. In a population of outbred rabbits, there are two different alleles of the gene that specifies H, and partially inbred strains that are homozygous for each of the alleles have been identified (53). The gene products of the two alleles (designated H_1 and H_2) differ in their apparent molecular weights by about 10,000 daltons, such that radiolabeled axonally transported H from the two strains of rabbits can be easily distinguished electrophoretically (Figure 4-2). When neurofilaments are partially purified from the two strains of rabbits, the high-molecular-weight polypeptide from the two preparations differs in electrophoretic mobility in the same strain-dependent manner as axonally transported H, indicating that they are products of the same gene. Furthermore, antibodies prepared against purified H (again identified by its strain-dependent electrophoretic mobility) react with neurofilaments; the reaction has been observed both as the ability of the antibody to mediate the adsorption of neurofilaments to "immunoaffinity" electron-microscope grids that had been coated with the antibody (97) and by direct electron-microscopic observation of the attachment of the antibodies to partially purified preparations of neurofilaments in negatively stained preparations (96) (Figure 4-3). These observations indicate that antigens that are specified by the gene coding for the group V H polypeptide are physically associated with neurofilaments.

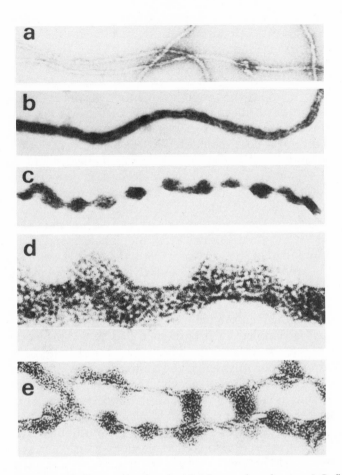

Figure 4-3. Purified neurofilaments incubated with (a) control nonimmune IgG; (b) anti-L; (c, d, and e) anti-H (a, b, c, and e are at similar magnifications; d is at higher magnification). Anti-L attaches in a continuous manner along the length of the filament (b), while anti-H attaches periodically (c). Sometimes the antigens that react with anti-H appear to compose a structure that helically wraps a nonreactive central core (d); occasionally, anti-H reactive material appears to bridge two filaments (e).

Several conclusions about the organization of the three triplet polypeptides within filaments have been generated by electron-microscopic observations of the attachment to neurofilaments of antibodies that react preferentially with H, M, or L (Figure 4-3). First, because all three antibodies react with most of the neurofilaments in purified preparations from spinal cord (77,86) or in cultured dorsal root ganglion neurons (82), a single filament can contain all three polypeptides; it is, therefore, unlikely that there are three classes of neurofilaments, each composed of a single species of triplet protein.

Second, anti-L antibodies appear to attach uniformly along the length of a neurofilament. In contrast, anti-H reacts with antigens that are periodically arranged along the filament (77,96). In some cases, the anti-H-decorated structure appears to be helically wrapped around an unreactive central core; the helix has a period of about 1000 Å and a unit length of about 1.5 periods. In other instances, anti-H reacts with material that bridges two filaments that are as far apart as 700 Å (96) (Figure 4-3). These observations suggest that neurofilaments comprise a central core, composed at least in part of L, and a more peripheral and loosely attached structure, composed in part of H (Figure 4-4); observations of anti-M-decorated neurofilaments suggest that M may also contribute to a peripheral structure, but its disposition within the filament is the least clear.

Although the interpretation of antibody-decorated filaments is subject to several uncertainties (96), the conclusion that L is an integral component of the filament core and that H contributes to a structure peripherally and loosely attached is consistent with recent observations that filaments can be assembled from purified L polypeptide but not from purified H alone (24,54,56,100). According to one report, filaments assembled in the absence of H appeared smooth, while those assembled in its presence had wispy projections from their surfaces (24). An additional indication that H may be tenuously associated with neurofilaments, and perhaps not required for the filamentous morphology, is that certain areas of the nervous system may contain filaments lacking H. The dendrites of cortical and hippocampal pyramidal cells contain intermediate filaments and can be immunofluorescently stained with anti-M and anti-L antibodies but not with anti-H (83). Furthermore, in certain developing neuronal systems, the appearance of H lags behind the appearance of M and L, suggesting

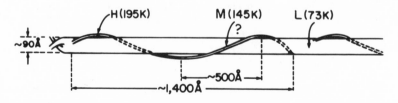

Figure 4-4. Schematic representation of the relationship of H, M, and L suggested by the appearance of antibody-decorated neurofilaments in Figure 4-3. Filaments comprise a central core, which contains L, and a structure that is loosely and peripherally attached to the central core that contains H and possibly M antigens. The helix shown here is one configuration of the H-containing peripheral structure that has been observed in antibody-decorated preparations *in vitro* (Figure 4-3[d]). Another configuration is illustrated in Figure 4-3(e), where it appears that the H-containing peripheral structure can project from the surface of the filament and form cross-bridges of the kind seen in Figure 4-1.

that there may be a period of time when neurofilaments lacking H are present in these neurons (45).

These indications that H is a component of a peripheral structure that is not an obligatory constituent of a neurofilament raises the possibility that H is an accessory neurofilament-associated protein (NAP), rather than an integral subunit of neurofilaments. The analogy to the high-molecular-weight MAPs is intriguing, because, as in the case of H, the high-molecular-weight MAPs appear to be exclusively neuronal; in addition, the high-molecular-weight MAPs are arranged periodically around the outisde of microtubules (15,40). In certain neurons (e.g., cortical and hippocampal pyramidal cells) where H appears to be most concentrated in the axons, high-molecular-weight MAPs are found only in the cell bodies and dendrites; the possibility that the complementary distribution of these two polypeptides reflects complementary functions has been noted (83).

On the other hand, certain observations suggest that, although a portion of H is peripheral to the central core of the filament and is not required for filament formation, H is nevertheless an integral subunit of neurofilaments. For example, H has only been dissociated from neurofilaments under conditions (e.g., 8-M urea) that denature proteins and disrupt the filaments, suggesting that the association between H and neurofilaments is intimate. Furthermore, immunological evidence indicates that H is structurally related to the polypeptides that form the central core of neurofilaments. Polyclonal antibodies raised against any one of the three filament polypeptides cross-react with the other two—in the case of one antiserum against L, it was estimated that as much as 60% of the antibodies cross-reacted with H (96)—suggesting that they are related by virtue of shared antigenic determinants. The additional observation that a certain monoclonal antibody reacts with all three neurofilament polypeptides (as well as with all other intermediate filament polypeptides) (68) confirms that at least one filament-specific antigen is shared by the three polypeptides. These relationships cannot be easily explained in terms of a common origin of the three polypeptides as posttranslational cleavage products of a single gene product, because all three polypeptides can be synthesized in a cell-free translation system, apparently from separate mRNA molecules (13). The antigenic relationships are, however, consistent with the possibility that the three polypeptides are specified by three genes that have evolved from a common ancestral gene (96).

The NAP-like properties of H (i.e., its peripheral disposition and its optional role in filament formation) and the properties suggesting that H is an integral subunit of neurofilaments (i.e., its intimate association with the filament and its structural relationship to core polypeptides) could both be harmoniously accommodated by a general model of intermediate filament structure originally proposed by Steinert and others (87,88) to explain the structure of epidermal intermediate filaments (keratin filaments) and the intermediate filaments of fibroblasts. According to this model, intermediate filaments are constructed from units consisting of three subunit polypeptide chains. Each chain contains two α-helical regions of invariant length (about 180 Å) separated by a nonhelical region of variable length depending on the length of the particular polypeptide. The three polypeptide chains are arranged within the unit in such a way that the α-helical regions of each polypeptide are in register with the α-helical regions of the other two. The α-helical regions of the different polypeptides are twisted around each other, forming two three-stranded helices that are separated and flanked by three strands of nonhelical polypeptide chains (Figure 4-5A). The unit, which is 20 Å in diameter and about 480 Å long, is combined with other units, side by side and end to end, to produce a complete filament. If neurofilaments were constucted from such units, each comprising an H, M, and L polypeptide (Figure 4-5A), the NAP-like and integral properties of H could be understood in the following way. First, the antigenic relationship of the three polypeptides would be a consequence of a requirement of homology between the matching α-helical portions of each of the three polypeptide chains. Second, the nonhelical region of H would be much longer than the corresponding regions of the other two polypeptides, and the excess H would be expected to project from the surface of the filament (Figure 4-5B). If the three-stranded units were arranged in the filament in such a way that the excess H projected at only one point in the span of a single unit, antibodies against H would be expected to attach to the filament preferentially at these projection sites. The anticipated distance between projections of about 480 Å (the length of one unit in Steinert and others' model) is similar to the observed distance between anti-H reactive nodes (about 500 Å) (96). The continuously helical appearance of the anti-H-decorated structure could be generated if one projection of H were linked to the next, either by virtue of a physiologically meaningful communication between sequential projections or by virtue of a crosslinking of material from adjacent projection sites by the divalent anti-H antibody. The absence of H from some areas of the nervous system would be compatible with this model if, like the intermediate keratin filaments of skin, several combinations of subunits could form filaments (e.g., L_3 or L_2M_1 as well as $L_1M_1H_1$).

Regardless of whether H is a peripherally attached NAP or an integral subunit with a portion projecting from the surface, it contributes to a structure peripheral to the central core of the filament, suggesting that its function may involve mediation of interactions between the filament and its environment. In this capacity, H is a candidate for the cross-bridges between neurofilaments and between neurofilaments and other organelles (Figure 4-1); this possibility is enhanced by the observation that the anti-H antibody can react with material that interconnects purified neurofilaments (Figure 4-3) and has been confirmed by the recent observation that anti-H can decorate these cross-bridges *in situ* (36). The helical configuration of antibody-decorated H might represent either the collapse of a cross-bridge onto the filament when the matrix is disrupted during filament purification or, if the cross-bridge is a dynamic structure that attaches and detaches from other filaments and organelles, the helix could represent a physiologically meaningful "detached" state. The helical configuration of antibody-decorated H wrapping a central core is reminiscent of tropomyosin, which is helically related to a central actin filament; like tropomyosin, which regulates interactions between actin and myosin, H, one could postulate, may regulate interactions between the central core and other elements of the axoplasm—both filaments and membrane-bounded organelles.

Molecular Form of Moving Neurofilament Polypeptides

Radiolabeled, newly synthesized neurofilament polypeptides move down axons slowly, in the company of tubulin (37,44), the microtubule-associated tau factors (3), and fodrin (47,95) (Figure 4-2). In the phrenic nerve of the guinea pig, the label associated with these group V proteins does not diminish appreciably as the proteins pass down the axon, indicating that they are extremely stable in transit (43). It has also been noted that the length of nerve occupied by the labeled proteins does not increase appreciably as they traverse the axon (37,43,44), and this observation has been interpreted to indicate that the group V polypeptides are restricted from altering their relationships to each other and from spreading along the length of the axon. This coherent movement of the newly synthesized group V polypeptides has suggested that, immediately after they are synthesized by ribosomes that may be already bound to the cytoskeleton (42,67), the neurofilament proteins polymerize and move down the axon in the form of intact neurofilaments cross-linked to each other and to microtubules (3,37,44). Although individual neurofilaments and microtubules are much shorter than the length of many axons (6,8,90) (estimates of average microtubule length have ranged between 108 μm [6] and 370 to 760 μm [90] in mammalian axons), the cross-linkers between them may generate a neurofilamentous microtubule lattice (NF-MT-lattice) that is continuous from the cell body

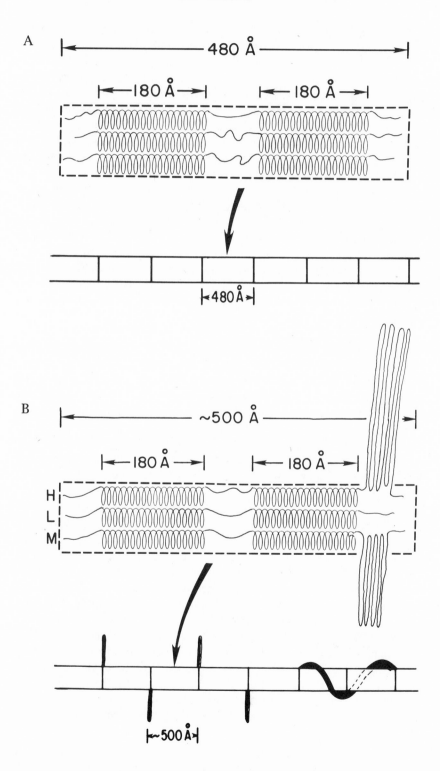

to the terminal regions of the axon. If the lattice were continuously elaborated from newly synthesized group V proteins in the soma and disassembled in the terminals, its continuous growth down the axon could account for the transport behavior of the group V polypeptides; because they would be immobilized in the same macromolecular structure, pulse-labeled group V polypeptides would retain their associations with each other and resist spreading during their journey down the axon. (The proteolytic disassembly of the neurofilamentous cytoskeleton in a restricted region near the synaptic terminal has been suggested by the observation that the triplet polypeptides, which appear stable during their passage down the axon, do not accumulate in synaptic terminals [42,65]. Several proteases with certain properties appropriate for such a terminal disassembly process have been described. The possibility of terminal disassembly and alternative possibilities for the proteolytic processing and catabolism of neurofilaments are considered in detail by R. Nixon in Chapter 5 of this volume and in reference 59.)

The existence of an NF-MT lattice is evident from the electron-microscopic observations that neurofilaments and microtubules interact via cross-bridges (18,34,89) (Figure 4-1); biochemical evidence has suggested that these interactions may be mediated in part by high-molecular-weight MAPs and the microtubule-associated tau factors (84). The possibility that the NF-MT lattice moves down the axon is consistent with the following observations: (1) filaments accumulate in axons when transport is blocked by chronic constriction of a nerve (91); (2) autoradiographically detected silver grains generated by labeled group V proteins occupy regions of the axon populated by

Figure 4-5. (A) model proposed by Steinert and others (87,88) originally to describe the relationship of polypeptides within keratin intermediate filaments. Filaments are constructed from units comprising three subunit polypeptide chains (of similar but not necessarily identical length), each of which contains two α-helical regions (coils) in register with the α-helical regions of the other two polypeptides. These α-helical regions from the three polypeptides are helically wrapped around each other (not shown) to form two three-stranded coils that are separated and flanked by short non-α-helical regions. The resulting unit is combined with similar units to form the intermediate filament. (B) Possible consequences of arranging H, M, and L into a unit of the model of Steinert and others shown in A. The length of H and M that is in excess of the length of L would be expected to project from the unit (upper part of B). The projecting portion of H is drawn at the end of the unit rather than between the two α-helical regions because of evidence that an end piece of H (molecular weight 160,000) can be proteolytically cleaved from intact neurofilaments without appreciably altering the appearance of the filament (Peter Eagles, personal communication). If the units within the filament are in register (lower left), the interval between the projections of the excess length of H and M would be the length of a unit (about 500 Å). This interval is similar to the interval between sites on neurofilaments that are most reactive with anti-H antibodies (Figure 4-3). These projections of H might assume a helical appearance if one projection were linked to the next (lower right), either in a physiologically meaningful way or by the divalent antibodies used to visualize H.

microtubules and neurofilaments (16); and (3) when nerves are exposed to β-β'-iminodipropionitrile (IDPN), a chemical that blocks the axonal transport of triplet polypeptides (32,99), neurofilaments accumulate in axon swellings close to the cell body (10,11).

However, there is no direct evidence that group V polypeptides are associated with the NF-MT lattice while they are moving, and the alternative hypothesis, that newly synthesized neurofilament polypeptides move down the axon as precursors to a stationary NF-MT lattice, is also compatible with most of the behavior of moving neurofilament proteins. For example, the similarity in movement of the different group V polypeptides could reflect either an association between the moving precursors of filaments and tubules or a similar mode of transport of the different group V polypeptides. The observation that is most difficult to explain in terms of such a moving-precursor hypothesis is the conservation of the labeled, newly synthesized group V polypeptides as they move down the axons of guinea pig phrenic nerve (43); if precursors were added to a stationary NF-MT lattice along the length of the axon, the amount of moving precursor would be expected to decrease along the length of the axon. To explain the conservation of group V-associated label, it would be necessary to suppose that only a small fraction of the moving precursor is added to the NF-MT lattice, which itself is very stable. Such a situation, in which most of the transported neurofilament precursor would reach the end of the axon unused, appears wasteful. However, the moving-precursor hypothesis cannot be dismissed on these grounds. It could be advantageous for the neuron to maintain a sufficient supply of precursor materials for major repairs in the eventuality of catastrophic injury to the NF-MT lattice even if only a small fraction of the material were required under normal circumstances. Alternatively, the possibility has not been excluded that group V polypeptides that are not incorporated into the NF-MT lattice perform some other function as they pass down the axon.

Certain considerations suggest that, if a moving lattice is the vehicle that conveys group V polypeptides down the axon, the lattice must have a considerable capacity for rearrangement of its elements. For example, in myelinated axons, the number of filaments is not constant along the axon's length; the number is markedly reduced at the unmyelinated nodes of Ranvier compared to the internodal regions (90). If the entire lattice moves as a unit, it must reorganize at the nodes and the velocity of neurofilament transport must increase, just as the velocity of water in a river increases as the river narrows. Furthermore, the relationship of the neuorfilaments to microtubules may be adjusted at the nodes, since the number of microtubules does not decrease appreciably in the nodal region (90). A considerable

degree of plasticity would be required of a moving lattice to accommodate the requirements of passing through a node, and it would be
anticipated that this plasticity would manifest itself in an alteration
in the relationship of individual group V polypeptides to each other
as they pass down the axon. Alternatively, if the matrix were stationary, it could be maintained by equilibration with a moving fraction
of precursors that would progress down the axon and through the
nodes at a constant velocity.

A second problem that a moving lattice might encounter is the repair of filaments damaged in transit. In the absence of a local repair
mechanism, it would be necessary for a filament to remain intact for
the entire time required to traverse the axon—possibly as long as
several years in long nerves of large mammals. However, if a coherent
column of moving lattice were sufficiently plastic, it might have the
capacity to repair damage adequately by local rearrangements of its
components. Alternatively, if a stationary lattice were continually
renewed by moving precursors, repair would be easily accomplished.

Both of these considerations suggest that, if group V polypeptides
are transported in association with a moving lattice, the lattice must
be plastic. It has been suggested, based upon the relative ease of extraction of tubulin and neurofilament proteins from axons (58), that
microtubules are labile, plastic structures in equilibrium with substantial amounts of unpolymerized tubulin, while neurofilaments are
extremely stable structures, whose polymerization is essentially irre-
versible. However, the filament components must change their relationship to each other as well as to microtubules as they pass through
the nodes of Ranvier and might be most resilient to injury if their
polymerization within the axon were reversible. Furthermore, unlike
certain other intermediate filaments, isolated mammalian neurofilaments are labile under conditions of low ionic strength (51,75), suggesting that appropriate alterations of local conditions within axons
could effect a plastic reordering of the neurofilamentous cytoskeleton.

The plasticity required of a moving NF-MT lattice would alter the
relationships of newly synthesized group V proteins (e.g., neurofilament proteins and tubulin) to each other as they pass down the axon,
and it would be expected that the labeled proteins would spread out
along the axon as they dissociated from each other. In addition,
tubulin would spread to a greater extent than the triplet if, as has been
suggested (42,58), tubulin exchanges more readily than the triplet
between a moving lattice and a pool of stationary precursors. These
expectations appear to be contradicted by the observation that group
V polypeptides move down the axon as a coherent wave; it has also
been suggested that the coherency of the wave is incompatible with
the idea that group V polypeptides are conveyed as moving precursors,

which would be expected to diffuse in transit. It is, therefore, worth considering the degree of dissociation that could be accommodated by the experimental evidence for coherent movements of group V material. The most convincing report that labeled group V proteins occupy a constant length of axons and are transported without decrement is the observation that a peak of group V-associated radioactivity has a similar size and shape in the guinea pig phrenic nerve at 32 and 64 days after the cell bodies of motor neurons in the spinal cord have been labeled (43). In other systems, a much greater degree of spreading has been observed, but this could reflect different transport velocities of group V proteins moving coherently in different axons of a heterogeneous population of axons. In the C fibers of the garfish olfactory nerve, a homogeneous population of axons, the slowly moving peak of radioactivity is not coherent as it moves down the axon; it spreads and declines (7). However, recent analysis of the labeled polypeptides that give rise to this peak indicates that they may not be typical group V polypeptides (P. Cancalon, personal communication), and their behavior may not reflect the behavior of neurofilament proteins. However, even in the cases where triplet polypeptides appear to move coherently (43), the length of the nerve segment that they occupy (at least 30 mm in the guinea pig phrenic nerve) is longer than would be anticipated if the proteins had been pulse labeled and immediately added to the end of a growing lattice destined for exit from the cell body. For example, if label were available for incorporation for a period of 2.4 hours, the pulse-labeled group V proteins should occupy only 0.1 mm of individual axons if they were transported at 1 mm per day. The observed width of the group V peak could be generated if, contrary to the idea that they immediately add to the growing lattice, newly synthesized group V polypeptides entered a large pool of precursors, from which filaments and tubules for transport are withdrawn at random; or, alternatively, if the population of cell bodies that were labeled varied in their distance from the analyzed segments of the nerve, so that the polypeptides at the front of the wave had traveled farther (about 30 mm in the guinea pig phrenic nerve) than the polypeptides in the rear. If, within a single axon, the pulse-labeled group V polypeptides that have recently been added to a growing lattice in fact occupy a length of axon that is only about 0.3% (0.1 mm) of the width of the wave generated by all the axons, the group V proteins within an axon could spread to occupy 30 times their initial length, and the width of the multiaxonal wave would increase by only about 10%—an increase that would be well within the reported measurements of the coherency of the wave. These considerations emphasize that the degree to which newly synthesized polypeptides remain associated as they pass down the axon is not precisely known; in the absence of this knowledge, the plasticity required of a moving lattice would not necessarily be in violation of the experimental evidence. The idea of a moving lattice is thus an attractive hypothesis for explaining the behavior of group V polypeptides but remains to be demonstrated definitively. On the other hand, because of the uncertainty in the actual degree of association between moving group V polypeptides, the degree that moving precursors would spread could also be within the limits imposed by the experimental observation of coherently moving peaks of group V polypeptides, and the possibility that the lattice is maintained by moving precursors cannot yet be ruled out.

Mechanism of Transport of Neurofilament Proteins

The mechanism that moves the triplet polypeptides down the axon is as much a mystery as the mechanism that moves more rapidly transported materials. Because of the difficulties associated with mainpulating the milieu of a neuron for a sufficient period of time to observe the effect upon slow transport, the parameters that influence the movement of the triplet have been studied less extensively than the parameters that influence the rapidly transported proteins. However, certain observations provide potential clues.

First, the velocity of movement of the triplet polypeptides can differ significantly in different neurons and even in two axon branches of the same neuron; for example, the velocity is about 0.25 mm per day and 1 mm per day in guinea pig (3,47) and rabbit (95) retinal ganglion cells, respectively, and 0.3 mm per day and 1 mm per day in the central and peripheral processes of rat dorsal root ganglion cells, respectively (57). However, the factors reponsible for these differences are not yet clear.

Second, the transport of the triplet proteins can be selectively blocked by IDPN (32,99), a compound that produces a peripheral neuropathy (10,11). When IDPN is administered systemically, polypeptides that had progressed a considerable distance down the axon before the initiation of IDPN intoxication are arrested, indicating that triplet transport is disrupted along the entire length of the axon (32). The transport of tubulin and actin (a group IV protein) can also be retarded by IDPN but to a lesser extent than the triplet. The movement of group I material is not affected (32). A characteristic of IDPN intoxication is the appearance of large, neurofilament-filled swellings in the axon close to the cell body. These may be generated by the continued supply of neurofilaments from the cell body to an axon that is unable to transport them (41).

Because IDPN blocks transport of the triplet selectively, its mode of action should provide clues to the mechanism of triplet transport. Its site of action appears to be the axon itself, because application of IDPN directly to the sciatic nerve produces a reversible local disruption of triplet transport accompanied by a local accumulation of neurofilaments. In the axons of IDPN-intoxicated animals, neurofilaments are displaced toward the periphery while microtubules occupy the central region; this suggests that IDPN uncouples the cross-bridges that normally interconnect the microtubules and neurofilaments (65). The molecular species that is the relevant target for IDPN is not yet known; its identification may clarify the mechanism of triplet transport.

A third consideration is that neurofilament proteins appear to be transported differently in different inbred strains of rabbits (95). In strain X/J, labeled triplet proteins transported in retinal ganglion cell axons lag behind the triplet proteins of an outbred strain (New Zealand white) by a constant interval of about 5 mm, suggesting that their exit from the cell body is delayed by about 5 days. Because the transport velocity of the triplet within the axon is unaltered in strain X/J, the transport of neurofilament proteins from their site of synthesis to the axon can apparently be influenced separately from their transport within the axon. Strain X/J is homozygous for the allele of the H gene that specifies an abnormally small form of H (H_2, apparent molecular weight 185,000) (92) (Figure 4-2); it is, therefore, reasonable to wonder whether the delayed exit from the cell body is a consequence of a partial deficit in the function performed by H. However, the differences in triplet export could also be a consequence of other uncharacterized differences between the strains, and trivial explanations (e.g., differences in the intraocular axon lengths), although unlikely, have not been rigorously excluded.

A fourth potential clue to the mechanism of triplet transport is that fodrin, a protein comprising two high-molecular-weight polypeptides (250,000 and 240,000) (47,93,94) appears to move in association with slowly transported proteins, including those in group V. Although a portion of fodrin moves down the axon rapidly, at the velocity of group II, the intensity of labeling of fodrin in the axon increases when labeled group V polypeptides enter the axon (47,95); in addition, labeled fodrin is associated with material of greater density (as judged by its sedimentation on sucrose density gradients) at times when labeled group V polypeptides are in the axon that at earlier times when only more rapidly transported proteins are present (52). These observations suggest that newly synthesized fodrin becomes associated with several different organelles (including the group V neurofilaments or microtubules) that are subsequently transported down the axons at velocities ranging from those of group II to group

V (52). Because one feature that these different organelles have in common is their transport down the axon, it is interesting to consider whether fodrin could be involved in the transport process.

Fodrin, which appears to be related to erythrocyte spectrin and a protein (TW 260/240) of the terminal web of the intestinal brush border (26,27), has several properties that might be useful for a protein involved in a mechanico-chemical transduction system. All three of these proteins are rod-shaped molecules that bind to calmodulin and actin and can cross-link actin filaments *in vitro* (8,26, 27,28,48,69,93; R. Cheney, unpublished observations from this laboratory). Fodrin has been reported to stimulate the actomyosin ATPase *in vitro* (85). In the terminal web of the intestinal brush border, a fodrinlike protein serves as a multifunctional cross-linker, linking bundles f-actin filaments to other actin bundles, to the plasma membrane, to membrane-bound vesicles, and to intermediate filaments; certain vessicle-plasma membrane linkages may also be mediated by this protein (35). Fodrin is most concentrated in the region underlying the plasma membrane; as a consequence, immunofluorescently labeled fodrin resembles a cellular lining.

In addition to its intracellular movements in axonal transport, fodrin has been observed to move during the process of capping in lymphocytes (45,46,49). In capping, surface moieties that have been cross-linked by a multivalent ligand are actively transported to one pole of the cell. Fodrin (as well as actin, α-actinin, calmodulin, and, in some cases, myosin) redistributes on the underside of the plasma membrane and forms a subcap beneath the cap of the surface moiety.

Such behavior illustrates that fodrin can move beneath the plasma membrane of lymphocytes, and has generated the speculation that lymphocyte capping and certain aspects of axonal transport may be analogous events (46,49). Both lymphocyte capping and the axonal transport of group IV involve the movement of fodrin, actin, myosin, and calmodulin; in addition, the velocity of protein translocation is similar in the two processes (49). The observation that in lymphocyte capping fodrin is apparently translocated within the compartment beneath the plasma membrane suggests that both processes may depend upon a "mobile lining," comprising in part fodrin, actin, and myosin. Actin and myosin could participate in generating the force for the moving lining. In axons, this "mobile lining" would be in continuous movement down the axon at the velocity of group IV and would account for the observation that radiolabeled group IV proteins are most concentrated at the periphery of the axon, as determined by electron-microscopic autoradiography (21). Certain other organelles, including the group V proteins, could be transported by interacting with such a mobile lining. For example, fodrin might serve to link neurofilaments, microtubules, or both to this peripheral force-generating apparatus; the translational force imparted to these peripheral tubules or filaments would then be transmitted to the rest of the lattice by the cross-bridges. These possibilities, although speculative, provide one framework for interpreting certain events that occur during the developmental generation of the neurofilamentous cytoskeleton, as discussed below.

Generation of the Neurofilamentous Cytoskeleton During Neuronal Development

The neurofilamentous cytoskeleton is a late-developing structure in certain neuronal systems. In rat optic nerve, neurofilaments do not appear in the retinal ganglion cell axons until about 5 days after birth; their numbers increase from this time into adulthood, and their spacing is reported to become more regular as the animal matures (66). The late appearance of filaments in mammalian retinal ganglion cell axons is paralleled by a late appearance of the triplet polypeptides. In optic nerves of neonatal rabbits, the triplet polypeptides cannot be detected on Coomassie-blue-stained polyacrylamide gels until 6 days after birth. From 6 to 18 days, M and L can be detected, but H does not accumulate sufficiently to be stained until after day 18 (45). (Silver stains and antibody stains, which can be more sensitive than Coomassie blue stains for detecting proteins on polyacrylamide gels, indicate that small amounts of M and L are present in the rabbit optic nerve by the first postnatal day and that H is present by 8 days.) A similar late and staggered accumulation of the triplet polypeptides has been observed in the rat visual system; in this case, the appearance of M is reported to precede the appearance of L (64). The sequential accumulation of M, L, and H may be a reflection of a sequential onset of their axonal transport. In the retinal ganglion cells of the rabbit, axonally transported M and L can be radiolabeled in the cell bodies on the first postnatal day, but transported H cannot be labeled until after day 6. During the course of neuronal development, the velocity of slowly transported polypeptides in the rabbit optic nerve decreases. Prior to postnatal day 18, the three triplet polypeptides are transported at a velocity (8 mm per day) that is indistinguishable from the velocity of the group IV proteins in these axons. During the third week after birth, at about the time that H can first be stained with Coomassie blue, the velocity of the triplet proteins decreases approximately eightfold; the velocity of the group IV polypeptides also slows to the adult rate of about 3 mm per day during this period (45). This reduction in the velocity of group IV provides an explanation for a previous report that the velocity of slowly moving radioactivity decreases by a factor of two during the third postnatal week in rabbit retinal ganglion cells (33); the more pronounced slowing of group V was not apparent in these previous experiments because most of the slowly transported label is associated with group IV in these neurons.

The sequential accumulation of the triplet polypeptides suggests that the generation of the neurofilamentous cytoskeleton during development occurs in several distinct stages and may involve special mechanisms different from those that maintain the cytoskeleton in mature neurons. In an initial stage, the M and L subunits would be transported, either in the form of H-less neurofilaments or as neurofilament precursors that would polymerize to form H-less filaments at a subsequent stage. During these early stages, they would be transported as components of group IV into axons devoid of neurofilaments. In terms of the hypothetical mobile-lining model suggested in the previous section, the filaments or precursors would be moved by a group IV, peripheral force-generating mechanism.

The coincidence of the accumulation of H to a level where it is a major axonal protein and the slowing of groups IV and V suggest that these events may be related. Depending upon its functions, H could retard the transport of group V proteins in one of several ways. As a cross-bridge, its appearance could generate the NF-MT lattice from individual filaments and tubules. The reduction in velocity of triplet proteins might then represent the transition from the transport of individual filaments, or precursors, to the transport of a moving lattice (Figure 4-6). The additional drag that the lattice would impose upon the peripheral force-generating machinery might then explain the twofold reduction in the transport velocity of group IV (alternatively, the formation of the lattice could directly retard the passage of certain group IV proteins through the axoplasm). The rationale for such a sequence of events could be understood if a NF-MT lattice can only be transported when it extends through an appreciable length of the axon, in which case these special developmental mechanisms would be required initially to establish the lattice.

Or if H functions to regulate interactions between a filament and its environment, the developmental accumulation of H could retard group V transport by inhibiting interactions between filaments and elements necessary for their transport. For example, if the central core contained sites that reacted either directly with elements of the transport motor or with microtubules that, in turn, interacted with the transport motor, the inhibition of these sites by H would result in a reduction in the transport velocity of the triplet proteins.

The delayed appearance of H with respect to M and L raises the question of whether H combines with H-less filaments that are already in the axon or, alternatively, whether neonatal, H-less filaments are

A.

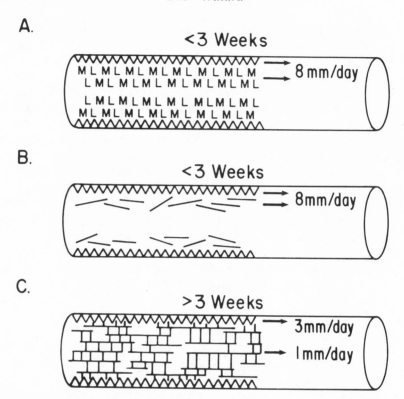

Figure 4-6. One hypothetical explanation for changes in transport rates during neonatal development in the rabbit visual system. Prior to 3 weeks of age, the neurofilament proteins M and L are transported as components of groups IV (sawtooth) in the form of unpolymerized precursors (A) or filaments lacking H (B). At about 3 weeks, H accumulates and cross-links the filaments. Henceforth, H, M, and L are transported as cross-linked filaments at group V velocities (C). The movement of group IV is retarded by the drag imposed upon it by the filament matrix.

replaced by adult filaments synthesized *de novo* in the cell body. The former alternative would be more easily accomplished if H is a peripherally associated NAP than if it were an integral polypeptide that had to exchange with an M or L polypeptide that was helically intertwined in the filament core. A third and less likely possibility is that the staggered appearance of the triplet polypeptides reflects different populations of axons: early-developing axons that would always, like certain dendrites, contain H-less filaments and a later-developing class that would contain a full complement of triplet proteins.

Do Neurofilaments Transport Other Organelles?

Although a structural role is the only function of neurofilaments that has yet been clearly demonstrated (25), certain properties of neurofilaments raise the suspicion that they may also participate in other, more dynamic processes such as axonal transport. For example, certain enzymatic activities, including a MAP-2-stimulated protein kinase (70), a cAMP-phosphodiesterase (71), and a Mg^{++}-ATPase (73,84), have been observed in preparations enriched with neurofilaments. In addition, certain aspects of neurofilament function may be regulated by posttranslational modifications, since all three of the triplet polypeptides of mammalian neurofilaments can be phosphorylated (38,74). (However, most of the Mg^{++}-ATPase in neurofilament preparations is not associated with the triplet polypeptides [29].)

The speculation that neurofilaments might be involved in the axonal transport of other materials is raised primarily by their arrangement in the axon; as long linear elements oriented in the same direction that materials are transported, they could provide tracks along which the transported materials could move. Furthermore, the cross-bridges between neurofilaments and intra-axonal organelles (e.g., vesicles and mitochondria) could reflect dynamic interactions responsible for propelling the organelles up or down the axon. However, the available evidence emphasizes the independence of axonal transport and neurofilaments.

For example, some axons, such as those in the ventral cord of the crayfish, do not contain neurofilaments (72). Crayfish axons can survive for long periods of time when separated from their cell bodies; this indicates that they are less dependent upon axonally transported materials than mammalian axons. However, within the mammalian nervous system, there is considerable variation in the numbers of neurofilaments in different types of axons, and it is not uncommon for neurofilaments to be undetectable morphologically and the triplet proteins undetectable by immunofluorescence (83). For example, neurofilaments have not been observed in electron micrographs of the unmyelinated axons of the vagus nerve, and the axonal transport of the labeled triplet polypeptides into the vagus nerve has not been observed after the vagal motor nuclei were radiolabeled with protein precurosors (99). Presumably, these neurons transport proteins by mechanisms not requiring neurofilaments.

The delayed appearance of neurofilaments during neuronal development also suggests that most aspects of axonal transport can be accomplished in the absence of neurofilaments, even in those neurons that will eventually contain them. The optic axons of newborn rats, which presumably depend upon axonal transport, lack neurofilaments (66). In 1-day-old rabbits, individual polypeptides representing all five adult transport groups are transported into the retinal ganglion cell axons (45); at this time, no H and only low levels of M and L can be detected by electrophoresis; this observation indicates that at least certain organelles of each transport group can move down the axon without the aid of neurofilaments.

The behavior of neurofilaments and rapidly transported materials during IDPN intoxication provides additional evidence that transport is independent of neurofilaments. In the axons of IDPN-intoxicated animals, the neurofilaments are displaced to the periphery of the axon. On the other hand, rapid transport not only persists but radio-labeled rapidly transported materials shift to the central portion of the axon, where they, along with mitochondria, smooth endoplasmic reticulum, and microtubules, appear to dissociate and segregate from the neurofilaments (21). In addition, when rapidly transported material is caused to accumulate by acute chilling of an axon, the neuro-filaments become compressed to the central region of the axon, and the organelles presumed to have been undergoing rapid transport accumulate in the periphery (89), consistent with the interpretation that they were transported independently of neurofilaments.

These observations provide a strong argument that neurofilaments are not required for many aspects of axonal transport. However, they do not definitively rule out the possibility that neurofilaments are necessary for transporting special materials required only by neurons that contain neurofilaments (e.g., adult but not immature retinal ganglion cell axons). Furthermore, the reduction in velocity of group IV proteins, which occurs during development at about the same time that the H polypeptide accumulates in neurons, suggests that filaments may influence the transport of other materials. In addition, the most rapidly moving radiolabeled materials in neonatal rabbit retinal ganglion cell axons have been reported to increase in velocity at about the same time that H accumulates (33), leaving open the possibility that neurofilaments may, directly or indirectly, facilitate the transport of group I.

Summary

An understanding of the relationship of neurofilaments to axonal transport—their own and the transport of other materials—will be most efficiently increased by learning the full repertoire of functional capabilities of neurofilaments and determining how each of the neurofilament's components contribute to these functions. The boundaries of understanding of the relationship of neurofilament structure to the axonal transport process are delineated by the following series of questions generated by the evidence that has been discussed here.

What is the precise relationship of the peripheral polypeptide H to the filament? Is it an accessory neurofilament protein (a NAP), or is it an integral protein, with a portion projecting away from the surface of the filament? Does it function to bridge neurofilaments with other organelles as well as with other neurofilaments? If so, does the bridge serve to translocate the organelle or to immobilize it? If not, what is the molecular nature of the bridging material? In addition to its role as an interfilamentous cross-bridge, can H regulate the interaction between filaments and other organelles by covering up reactive sites on the central core? What is the relationship of M to the filament? Is it a part of the L-containing core, the H-containing peripheral structure, or a third structure? What other elements are associated with neurofilaments?

In what form are filaments transported? As a polymerized NF-MT lattice or as precursors to the lattice? What is the motor for neurofilament transport? Is it at the periphery of the cell, utilizing peripheral fodrin, actin, and myosin? Do the peripherally displaced filaments in IDPN-intoxicated neurons represent a "rigor complex" of filaments with the transport motor or, conversely, an uncoupling of the filaments from the transport motor? With what molecules does IDPN react to cause its block of neurofilament transport?

What is the relevant difference between neurons that contain neurofilaments and those that do not? Is H universally associated with axonal neurofilaments but to a lesser degree with dendritic neurofilaments? Is the distribution of H and high-molecular-weight MAPs complementary within the nervous system? Do they perform variations of the same function? What is the relevant difference between axons that contain H and dendrites that contain less H?

Why does the appearance of M and L precede H in certain axons during development? Are there H-less filaments in neonatal axons, or are M and L present as precursors, awaiting H? What is the nature of the transition that occurs when H appears during development? Does it generate a NF-MT lattice where there was none before? What is the signal that initiates the transport of H and the accumulation of H? Could it be generated in postsynaptic cells and transported retrogradely to the cell bodies only when axons have reached their targets? Could the appearance of H stabilize cells that had made appropriate synaptic connections? Could the appearance of H be related to the period of cell death during development? For example, could H counteract the influence of factors that would otherwise destroy the neuron? Could the appearance of H be related in any way to the critical period of the visual system?

The potential for finding the answers to questions such as these should provide incentive for further explorations into the relationships between neurofilaments and axonal transport.

REFERENCES

1. Baitinger, C., J. Levine, T. Lorenz, C. Simon, P. Skene, and M. Willard. 1982. Characteristics of axonally transported proteins. *In*: Axoplasmic Transport. D. G. Weiss, editor. Springer-Verlag. Berlin, Heidelberg, pp. 111-120.

2. Black, M. M., and R. J. Lasek. 1979. Axonal transport of actin: Slow component b is the principal source of actin for the axon. *Brain Res. 171*:401-413.

3. Black, M. M., and R. J. Lasek. 1980. Slow component of axonal transport: Two cytoskeletal networks. *J. Cell Biol. 86*:616-623.

4. Brady, S. T., and R. J. Lasek. 1981. Nerve-specific enolase and creatine phosphokinase in axonal transport: Soluble proteins and the axoplasmic matrix. *Cell 23*:515-523.

5. Brady, S. T., M. Tytell, K. Heriot, and R. J. Lasek. 1981. Axonal transport of calmodulin: A physiologic approach to identification of long-term associations between proteins. *J. Cell Biol. 89*:607-614.

6. Bray, D., and M. B. Bunge. 1981. Serial analysis of microtubules in cultured rat sensory axons. *J. Neurocytol. 10*:589-605.

7. Cancalon, P. 1979. Influence of temperature on the velocity and on the isotope profile of slowly transported labeled proteins. *J. Neurochem. 32*:997-1007.

8. Carlin, R. K., D. C. Bartelt, and P. Siekevitz. 1983. Identification of fodrin as a major calmodulin binding protein in postsynaptic density preparations. *J. Cell Biol.*, in press.

9. Chalfie, M., and J. N. Thomson. 1979. Organization of neuronal microtubules in the nematode *Caenorhabditis elegans. J. Cell Biol. 82*:278-289.

10. Chou, S. M., and H. A. Hartman. 1965. Electron microscopy of focal neuroaxonal lesions produced by β, β'-iminodipropionitrile (IDPN) in rats: I. The advanced lesions. *Acta Neuropathol. 4*:590-603.

11. Chou, S. M., and R. A. Klein. 1972. Autoradiographic studies of protein turnover in motoneurons of IDPN-treated rats. *Acta Neuropathol. 22*:183-189.

12. Cooper, P. D., and R. S. Smith. 1974. The movement of optically detectable organelles in myelinated axons of *Xenopus laevis. J. Physiol. (Lond). 242*:77-97.

13. Czosnek, H., D. Soifer, and H. M. Wisniewski. 1980. Studies on the biosynthesis of neurofilament proteins. *J. Cell Biol. 85*:726-734.

14. Dahlstrom, A. 1967. The transport of noradrenaline between two simultaneously performed ligations of the sciatic nerves of rat and cat. *Acta Physiol. Scand. 69*:158-166.

15. Dentler, W. L., S. Garnett, and J. L. Rosenbaum. 1975. Ultrastructural localization of the high molecular weight proteins associated with *in vitro*-assembled brain microtubules. *J. Cell Biol. 65*:237-241.

16. Droz, B., H. L. Koenig, and L. DiGiamberardino. 1973. Axonal migration of protein and glycoprotein to nerve endings: I. Radioautographic analysis of the renewal of protein in nerve endings of chicken ciliary ganglion after intracerebral injection of [^{3}H] lysine. *Brain Res. 60*:93-127.

17. Edström, A., and M. Hanson. 1975. The mechanisms of fast axonal transport: A pharmacological approach. *Neuropharmacology 14*:181-188.

18. Ellisman, M. H., and K. R. Porter. 1980. Microtrabecular structure of the axoplasmic matrix: Visualization of cross-linking structures and their distribution. *J. Cell Biol. 87*: 464-479.

19. Erickson, P. F., K. B. Seamon, B. W. Moore, R. S. Lasher, and L. N. Minier. 1980. Axonal transport of the Ca^{2+}-dependent protein modulator of 3'-5'-cyclic-AMP phosphodiesterase in the rabbit visual system. *J. Neurochem. 35*:242-248.

20. Forman, D. S., A. L. Padjen, and G. R. Siggins. 1977. Axonal transport of organelles visualized by light microscopy: Cinemicrographic and computer analysis. *Brain Res. 136*: 197-213.

21. Gambetti, P. L., K. Heriot, and R. J. Lasek. 1982. Personal communication.

22. Garcia, A. G., S. M. Kirpekar, J. C. Prat, and A. R. Wakade. 1974. Metabolic and ionic requirements for the axoplasmic transport of dopamine beta-hydroxylase. *J. Physiol. (Lond.)*. *241*:809-821.

23. Garner, J., and R. J. Lasek. 1981. Clathrin is axonally transported as part of slow component b: The microfilament complex. *J. Cell Biol. 88*:172-178.

24. Giesler, N., and K. Weber. 1981. Self-assembly *in vitro* of the 68,000 molecular weight component of the mammalian neurofilament triplet proteins into intermediate-sized filaments. *J. Mol. Biol. 151*:565-571.

25. Gilbert, D. S. 1975. Axoplasm architecture and physical properties as seen in the *Myxicola* giant axon. *J. Physiol. (Lond.). 253*:257-301.

26. Glenney, J. R., P. Glenney, M. Osborn, and K. Weber. 1982. An F-actin-and-calmodulin binding protein from isolated intestinal brush borders has a morphology related to spectrin. *Cell 28*:843-854.

27. Glenney, J. R., P. Glenney, and K. Weber. 1982. Erythroid spectrin, brain fodrin, and intestinal brush border proteins (TW 260/240) are related molecules containing a common calmodulin-binding subunit bound to a variant cell type-specific subunit. *Proc. Natl. Acad. Sci. USA 79*:4002-4005.

28. Glenney, J. R., P. Glenney, and K. Weber. 1982. F-actin-binding and cross-linking properties of porcine brain fodrin, a spectrin-related molecule. *J. Biol. Chem 257*:9781-9787.

29. Glicksman, M. A., and M. Willard. 1982. Separation of neurofilaments from ATPase activity by precipitation with anti-neurofilament antibodies. *J. Neurochem. 38*:1774-1776.

30. Goldberg, D. J., D. A. Harris, B. W. Lubit, and J. H. Schwartz. 1980. Analysis of the mechanism of fast axonal transport by intracellular injection of potentially inhibitory macromolecules: Evidence for a possible role of actin filaments. *Proc. Natl. Acad. Sci. USA 77*:7448-7452.

31. Grafstein, B., and D. S. Forman. 1980. Intracellular transport in neurons. *Physiol Rev. 60*:1167-1283.

32. Griffin, J. W., P. N. Hoffman, A. W. Clark, P. T. Carroll, and D. L. Price. 1978. Slow axonal transport of neurofilament proteins: Impairment by β, β'-iminodipropionitrile administration. *Science 202*:633-635.

33. Hendrickson, A. E., and M. Cowan. 1971. Changes in the rate of axoplasmic transport during postnatal development of the rabbit's optic nerve and tract. *Exp. Neurol. 30*:403-422.

34. Hirokawa, N. 1982. The crosslinker system between neurofilaments, microtubules and membranous organelles in the axons revealed by quick freeze, deep etching method. *J. Cell Biol. 94*:129-142.

35. Hirokawa, N., R. Cheney, and M. Willard. 1983. Location of a protein of the fodrin-spectrin-TW 260/240 family in the mouse intestinal brush border. Manuscript submitted.

36. Hirokawa, N., M. A. Glicksman, and M. Willard. 1982. Organization of mammalian neurofilament polypeptides within the axonal cytoskeleton. *J. Cell. Biol. 95*:236 (abs.).

37. Hoffman, P. N., and R. J. Lasek. 1975. The slow component of axonal transport: Identification of major structural polypeptides of the axon and their generality among mammalian neurons. *J. Cell Biol. 66*:351-356.

38. Honchar, M. P., M. B. Bunge, and H. C. Agrawal. 1982. *In vivo* phosphorylation of NF proteins in the CNS of immature rat and rabbit. *Neurochem. Res. 7*:365-372.

39. Isenberg, G., P. Schubert, and G. W. Kreutzberg. 1980. Experimental approach to test the role of actin in axonal transport. *Brain Res. 194*:588-593.

40. Kim, H., L. I. Binder, and J. L. Rosenbaum. 1979. The periodic association of MAP-2 with brain microtubules *in vitro*. *J. Cell Biol. 80*:266-276.

41. Kirkpatrick, J. B., J. J. Bray, and S. M. Palmer. 1972. Visualization of axoplasmic flow *in vitro* by Nomarski microscopy. Comparison to rapid flow of radioactive proteins. *Brain Res. 43*:1-10.

42. Lasek, R. J. 1981. The dynamic ordering of neuronal cytoskeletons. *Neurosciences Res. Prog. Bull. 19*:7-32.

43. Lasek, R. J., and M. M. Black. 1977. How do axons stop growing? Some clues from the metabolism of the proteins in the slow component of axonal transport. *In*: Mechanisms, Regulation and Special Functions of Protein Synthesis in the Brain. S. Roberts, A. Lajtha, and W. H. Gispen, editors. Elsevier/North-Holland Biomedical Press, Amsterdam, pp. 161-169.

44. Lasek, R. J., and P. N. Hoffman. 1976. The neuronal cytoskeleton, axonal transport and axonal growth. *In* Cell Motility. Vol. 3. Microtubules and Related Proteins. R. Goldman, T. Pollard, J. Rosenbaum, editors. Cold Spring Harbor Laboratory, Cold Spring Harbor, N.Y., pp. 1021-1049.

45. Levine, J., C. Simon, and M. Willard, 1982. Mechanistic implications of the behavior of axonally transported proteins. *In*: Axoplasmic transport. D. G. Weiss, editor. Springer-Verlag. Berlin, Heidelberg, pp. 275-278.

46. Levine, J., J. H. P. Skene, and M. Willard. 1981. GAPs and fodrin: Novel axonally transported proteins. *Trends Neurosci. 4*:273-277.

47. Levine, J., and M. Willard. 1980. The composition and organization of axonally transported proteins in the retinal ganglion cells of the guinea pig. *Brain Res. 194*:137-154.

48. Levine, J., and M. Willard. 1981. Fodrin: Axonally transported polypeptides associated with the internal periphery of many cells. *J. Cell Biol. 90*:631-643.

49. Levine, J., and M. Willard. Redistribution of fodrin (a component of the cortical cytoplasm) accompanying capping of cell surface molecules. *Proc. Natl. Acad. Sci. USA*, in press.

50. Liem, R. K. H., and S. B. Hutchison. Purification of the individual components of the neurofilament triplet: Filament assembly from the 70,000 dalton subunit. *Biochemistry 21*:3221-3226.

51. Liem, R. K., S.-H. Yen, G. D. Salomon, and M. L. Shelanski. 1978. Intermediate filaments in nervous tissue. *J. Cell Biol. 79*:637-645.

52. Lorenz, T., and M. Willard. 1978. Subcellular fractionation of intra-axonally transported polypeptides in the rabbit visual system. *Proc. Natl. Acad. Sci. USA 75*:505-509.

53. Lubinska, L., and S. Niemierko. 1971. Velocity and intensity of bidirectional migration of acetylcholinesterase in transected nerves. *Brain Res. 27*:329-342.

54. Matus, A., R. Bernhardt, and T. Hugh-Jones. 1981. High molecular weight microtubule-associated proteins are preferentially associated with dendritic microtubules in brain. *Proc. Natl. Acad. Sci. USA 78*:3010-3014.

55. Metzuals, J., V. Montepetit, and D. F. Clapin. 1981. Organization of the neurofilamentous network. *Cell Tissue Res. 214*:455-482.

56. Moon, H. M., T. Wisniewski, P. Merz, J. DeMartini, and H. M. Wisniewski. 1981. Partial purification of neurofilament subunits from bovine brains and studies on neurofilament assembly. *J. Cell Biol. 89*:560-567.

57. Mori, H., Y. Komiya, and M. Kurokawa. 1979. Slowly migrating axonal polypeptides: Inequalities in their rate and amount of transport between two branches of bifurcating axons. *J. Cell Biol. 82*:174-184.

58. Morris, J. R., and R. J. Lasek. 1982. Stable polymers of the axonal cytoskeleton: The axoplasmic ghost. *J. Cell Biol. 92*:192-199.

59. Nixon, R. A., B. A. Brown, and C. A. Marotta. 1982. Post-translational modification of a neurofilament protein during axoplasmic transport: Implications for regional specialization of CNS axons. *J. Cell Biol. 94*:150-158.

60. Norton, W. T., and J. E. Goldman. 1980. Neurofilaments. *In*: Proteins of the Nervous System. 2nd ed. R. A. Bradshaw and D. M. Schneider, editors. Raven Press, New York, pp. 301-329.

61. Ochs, S. 1971. Local supply of energy to the fast axoplasmic transport mechanism. *Proc. Natl. Acad. Sci. USA 68*:1279-1282.

62. Ochs, S., and D. Hollingsworth. 1971. Dependence of fast axoplasmic transport in nerve on oxidative metabolism. *J. Neurochem. 18*:107-114.

63. Ochs, S., and N. Ranish. 1969. Characteristics of the fast transport system in mammalian nerve fibers. *J. Neurobiol. 1*:247-261.

64. Pachter, J. S., and R. K. H. Liem. 1981. Differential appearance of neurofilament triplet polypeptides in developing rat optic nerve. *J. Cell Biol. 91*:86 (abs.).

65. Papasozomenos, S. C., L. Autilio-Gambetti, and P. Gambetti. 1981. Reorganization of axoplasmic organelles following β, β'-iminodipropionitrile administration. *J. Cell Biol. 91*: 866-871.

66. Peters, A., and J. E. Vaughn. 1967. Microtubules and filaments in the axons and astrocytes of early postnatal rat optic nerves. *J. Cell Biol. 32*:113-119.

67. Porter, K. R., H. R. Byers, and M. H. Ellisman. 1979. The cytoskeleton. *In*: The neurosciences: Fourth Study Program. F. O. Schmitt and F. G. Worden, editors. MIT Press, Cambridge, Mass., pp. 703-722.

68. Pruss, R. M., R. Mirsky, M. D. Raff, R. Thorpe, A. J. Dowding, and B. H. Anderton. 1981. All classes of intermediate filaments share a common antigenic determinant defined by a monoclonal antibody. *Cell 27*:419-428.

69. Repasky, E. G., B. L. Granger, and E. Lazarides. 1982. Widespread occurrence of avian spectrin in non-erythroid cells. *Cell 29*:821-833.

70. Runge, M. S., M. R. El-Maghrabi, T. H. Claus, S. J. Pilkis, and R. C. Williams, Jr. 1981. A MAP-2 stimulated protein kinase activity associated with neurofilaments. *Biochemistry 20*:175-180.

71. Runge, M. S., P. B. Hewgley, D. Puett, and R. C. Williams, Jr. 1979. Cyclic nucleotide phosphodiesterase activity in 10 nm filaments and microtubule preparations from bovine brain. *Proc. Natl. Acad. Sci. USA 76*:2561-2566.

72. Samson, F. E. 1971. Mechanism of axoplasmic transport. *J. Neurobiol. 2*:347-360.

73. Schecket, G., and R. J. Lasek. 1977. A neurofilament associated Mg^{++}/Ca^{++} ATPase from squid giant axon. *Trans Am. Soc. Neurochem. 8*:179.

74. Schecket, G., and R. J. Lasek. 1979. Phosphorylation of neurofilaments from mammalian peripheral nerve. *Trans Am. Soc. Neurochem. 10*:140.

75. Schlaepfer, W. W. 1977. Immunological and ultrastructural studies of neurofilaments isolated from rat peripheral nerve. *J. Cell Biol. 74*:226-240.

76. Schlaepfer, W. W., and L. A. Freeman. 1978. Polypeptides of neurofilaments isolated from rat peripheral and central nervous systems. *J. Cell Biol. 78*:653-662.

77. Schlaepfer, W. W., V. Lee, and H. L. Wu. 1981. Assessment of immunological properties of neurofilament triplet proteins. *Brain Res. 226*:259-272.

78. Schlaepfer, W. W., and S. Micko. 1978. Chemical and structural changes of neurofilaments in transected rat sciatic nerves. *J. Cell Biol. 78*:369-378.

79. Schnapp, B. J., and T. S. Reese. 1982. Cytoplasmic structure in rapid frozen axons. *J. Cell Biol. 94*:667-679.

80. Schwartz, J. H. 1979. Axonal transport: Components, mechanisms, and specificity. *Ann. Rev. Neurosci. 2*:467-504.

81. Selkoe, D. J., R. K. H. Liem, S.-H. Yen, and M. L. Shelanski. 1979. Biochemical and immunological characterization of neurofilaments in experimental neurofibrillary degeneration induced by aluminum. *Brain Res. 163*:235-252.

82. Sharp, G. A., G. Shaw, and K. Weber. 1982. Immunoelectronmicroscopical localization of the three neurofilament triplet proteins along neurofilaments of cultured dorsal root ganglion neurons. *Exp. Cell Res. 137*:403-413.

83. Shaw, G., M. Osborn, and K. Weber. 1981. An immunofluorescence microscopical study of the neurofilament triplet proteins, vimentin, and glial fibrillary acidic protein within the adult rat brain. *Eur. J. Cell Biol. 26*:68-82.

84. Shelanski, M. L., J. F. Leterrier, and R. K. H. Liem. 1981. Evidence for interactions between neurofilaments and microtubules. *Neurosciences Res. Prog. Bull. 19*:32-43.

85. Shimo-Oka, T., and Y. Watanabe. 1981. Stimulation of actomyosin Mg^{2+}-ATPase activity by a brain microtuble-associated protein fraction: High-molecular weight actin-binding protein is the stimulating factor. *J. Biochem. 90*:1297-1307.

86. Stearns, M. E. 1982. High voltage electron microscopy of axoplasmic transport in neurons: A possible regulatory role for divalent cations. *J. Cell Biol. 92*:765-776.

87. Steinert, P. M. 1978. Structure of the three-chain unit of the bovine epidermal keratin filament. *J. Mol. Biol. 123*:49-70.

88. Steinert, P. M., W. W. Idler, and R. D. Goldman. 1980. Intermediate filaments of baby hamster kidney (BHK-21) cells and bovine epidermal keratinocytes have similar ultrastructural and subunit domain structures. *Proc. Natl. Acad. Sci. USA 77*:4534-4538.

89. Tsukita, S., and H. Ishikawa. 1980. The movement of membranous organelles in axons: Electron microscopic identification of anterogradely and retrogradely transported organelles. *J. Cell Biol. 84*:513-530.

90. Tsukita, S., and H. Ishikawa. 1981. The cytoskeleton in myelinated axons: Serial section study. *Biomed. Res. 2*:424-437.

91. Weiss, P., and H. B. Hiscoe. 1948. Experiments on the mechanism of nerve growth. *J. Exp. Zool. 107*:315-395.

92. Willard, M. B. 1976. Genetically determined protein polymorphism in the rabbit nervous system. *Proc. Natl. Acad. Sci. USA 73*:3641-3645.

93. Willard, M. 1977. The identification of two intra-axonally transported polypeptides resembling myosin in some respects in the rabbit visual system. *J. Cell Biol. 75*:1-11.

94. Willard, M., W. M. Cowan, and P. R. Vagelos. 1974. The polypeptide composition of intra-axonally transported proteins: Evidence for four transport velocities. *Proc. Natl. Acad. Sci. USA 73*:3641-3645.

95. Willard, M. B., and K. L. Hulebak. 1977. The intra-axonal transport of polypeptide H: Evidence for a fifth (very slow) group of transported proteins in the retinal ganglion cells of the rabbit. *Brain Res. 136*:289-306.

96. Willard, M., and C. Simon. 1981. Antibody decoration of neurofilaments. *J. Cell Biol. 89*:198-205.

97. Willard, M., C. Simon, C. Baitinger, J. Levine, and P. Skene. 1980. Association of an axonally transported polypeptide (H) with 100-Å filaments: Use of immunoaffinity electron microscope grids. *J. Cell Biol. 85*:587-596.

98. Willard, M., M. Wiseman, J. Levine, and P. Skene 1979. Axonal transport of actin in rabbit retinal ganglion cells. *J. Cell Biol. 81*:581-591.

99. Yokoyama, K., S. Tsukita, H. Ishikawa, and M. Kurokawa. 1980. Early changes in the neuronal cytoskeleton caused by β, β′-iminodipropionitrile: Selective impairment of neurofilament polypeptides. *Biomed. Res. 1*:537-547.

100. Zackroff, R. V., W. W. Idler, P. M. Steinert, and R. D. Goldman. 1982. *In vitro* reconstitution of intermediate filaments from mammalian triplet polypeptides. *Proc. Natl. Acad. Sci. USA 79*:754-757.

Chapter 5

Proteolysis
of Neurofilaments

Ralph A. Nixon

Proteolytic enzymes influence cell function at virtually every level of cellular organization. By relieving cells of obsolescent proteins and by replenishing supplies of free amino acids, proteinases exert coarse control over the entire cellular protein pool (1,94,135). Certain of these enzymes comprise a scavenging mechanism to protect cells from accumulating abnormal and potentially toxic proteins (41,72). An increasing number of proteinases with limited specificity has been shown to sculpture proteins in ways that modify their function or release smaller active polypeptides (reviews in 75,85,87,97,123). Limited proteolysis also may be the initial and rate-limiting step in the breakdown of most proteins (97). Indeed, this first proteolytic step seems to be more critical than synthesis rate in governing the steady-state level of certain proteins (5).

Each of these facets of proteolysis has potential relevance to how neurofilaments function in normal and pathological neurons. In this chapter, the evidence for metabolism of neurofilament proteins is reviewed within the broader context of proteolytic mechanisms known to exist in nervous tissue. Although this evidence is just beginning to unfold, the role of proteolysis in regulating neurofilament proteins (NFPs) appears to be an active and highly complex one.

I am grateful to L. I. Benowitz, D. P. Stromska, and B. A. Brown for their thoughtful comments on the manuscript. Supported by NIH grants NS15494, NS17535, and RR05404; the Sloan Foundation; and the Medical Foundation, Inc.

117

Methodologies for the Study of Neurofilament Proteolysis

The detailed examination of neurofilament proteolysis, as in other studies of proteolysis in nervous tissue, is subject to various methodological obstacles that are worth considering briefly. Very often, the actions of individual proteolytic enzymes have been studied by measuring hydrolytic activity in tissue homogenates or extracts. Tissue homogenization, however, disrupts the normal ionic environment, destroys proteolytic control mechanisms that depend upon the integrity of subcellular organelles (89,96,150,151), and alters topographical relationships among proteolytic enzymes and their substrates. Nonspecific proteinases that are normally compartmentalized may become exposed to proteins not encountered *in vivo*. Certain enzymes are rendered more unstable and may be subject to increased autolysis or inhibition by unphysiologically released inhibitors (50, 86). For example, when tissue integrity is disrupted, calcium-activated proteinases, which are active toward neurofilament proteins, may retain less than 10% of their original activity (R. A. Nixon, unpublished results; 84). This seems to be due in part to inhibition by ions (50) and endogenous inhibitors (26,95,144,152). Finally, the cellular heterogeneity of nervous tissue further complicates proteolytic measurements in most studies, since proteolytic activity in one cell type may be masked by the contributions from unrelated cells. Even when the enzyme and substrate are studied in purified form, certain of the foregoing considerations apply. (See the next section.)

Alternative strategies for assessing proteolysis, which emphasize the protein substrate rather than individual proteinase activities, focus on degradative rates of cellular proteins usually radiolabeled *in vivo* (reviews in 36,162). Recent adaptations of this approach (5,37, 45,112) circumvent the technical pitfalls in early studies (review in 47) and have permitted average turnover rates for brain proteins to be determined. Thus far, however, half-lives for only a few individual proteins in nervous tissue have been established (23,52,53,160). Neurofilament proteins are amenable to this *in vivo* analysis and, being neuron specific, may provide a window for examining neuronal proteolytic mechanisms selectively. Since half-lives obtained by current techniques reflect the average metabolism in all parts of neurons (e.g., perikaryon, axon, synapse, etc.), additional methods would be needed to examine whether neurofilament protein metabolism varies in different neuronal types or in different neuronal compartments within the same neuron. The contributions of individual enzymes to the overall metabolic process would also be difficult to discern by using only the existing *in vivo* methods.

Much of the present knowledge of neurofilament proteolysis has been derived from studies that strike a compromise between cell-free and purely *in vivo* approaches. In these studies, excised but otherwise intact segments of peripheral nerve or primary optic pathway are incubated *in vitro* under conditions that enhance or inhibit specific proteolytic enzymes (Chapter 3; 98,99,100,101,102,125,126,130). The time-dependent disappearance of individual proteins may then be measured through electrophoretic techniques. By partially preserving the cytotypic and subcellular integrity of the tissue (Figure 5-1) and, possibly, the specific topography of enzymes within their normal cellular compartments, one may obtain results that are especially relevant to *in vivo* processes. Since neurofilament proteins are contained only within axons, the loss of individual NFP species may be presumed to reflect neuronal proteolytic events. Furthermore, in the absence of neuronal cell bodies, new neurofilament protein synthesis is not a complicating factor in the analysis of proteolysis (10).

In a further adaptation of this approach, axonal and synaptic proteins may be specifically labeled *in vivo* prior to the *in vitro* analysis (98,99,100,101,102). For example, when radiolabeled amino acids

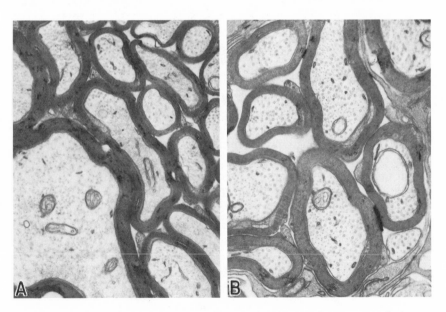

Figure 5-1. Electron-microscopic appearance of mouse optic nerve (transverse section): (A) before incubation and (B) after incubation at 37°C for 2 hours in HEPES medium pH 7.4 (containing 4-mM Ca^{++}) (98). Apart from some slight swelling of the myelin sheath, the integrity of organelles within axons and glia is largely preserved. (The excised nerves were fixed by immersion in 6% glutaraldehyde and postfixed in OsO$_4$. Magnification × 11,200.)

are administered intravitreally, labeled proteins synthesized in retinal ganglion cell (RGC) perikarya are conveyed by axoplasmic transport into the optic nerve and tract. The fate of labeled neurofilament proteins within different neuronal compartments may then be examined *in vivo* (101; see p. 139). In addition, enzymes comprising the proteolytic apparatus within axons and their activities toward individual axonal or synaptic proteins may be examined *in vitro* by incubating excised optic pathways under appropriate conditions (see p. 130). The axonal (neuronal) specificity of such *in vitro* measurements has been demonstrated (99).

In addition to being cell specific, this strategy makes it possible to detect proteinases that would be inactivated by tissue homogenization and to examine the actions of *individual* enzymes toward their endogenous substrates under environmental conditions that resemble the *in vivo* state. Indeed, the rate of protein degradation in RGC axons (98) or brain slices *in vitro* (36,37) approximates values obtained by *in vivo* techniques that take into account the equilibrium of precursor pools with label and the reutilization of labeled amino acids (5, 37,42,45). These approaches, however, are also subject to potential limitations. Thus far, studies employing this strategy have focused on neurofilament proteolysis within axons, which is not necessarily the major site of neurofilament metabolism. Since axons are separated from their cell bodies before proteolysis is measured *in vitro*, proteolytic activity may also partly reflect reactive responses of axons to axotomy. Even though several observations suggest that the magnitude of this problem is small (98), the relationship between protein-degradative rates in tissue slices or segments and rates *in vivo* should be cautiously interpreted. Furthermore, although altering ionic conditions and employing enzyme inhibitors *in vitro* can be a particularly valuable way of selectively enhancing individual proteinase activities, predicting the effect of these manipulations on the intracellular environment may be difficult. Precise definition of certain enzyme properties, therefore, requires correlative measurements on the purified enzyme. It is worth considering, however, that, since properties of certain enzymes may change during purification (61,50,71; also, see next section), the properties observed in the undisrupted tissue may sometimes reflect behavior of the enzyme *in vivo* more accurately than those observed when the enzyme is purified.

Proteinases in the Nervous System

The list of proteinases that have been identified in nervous tissue has expanded considerably in recent years. If the more advanced knowledge of nonneural systems serves as a precedent, however, the current

Table 5-1
Proteinases in Nervous Tissue

Enzyme (Trivial Name)	Catalytic Mechanism	pH Range for Activity	Activators	Inhibitors for Identification	Reference(s)
Calcium-activated neutral proteinase I	Cysteine (thiol)	6.0 to 9.0	Ca^{++} (mM), free sulfhydryl	EGTA, iodoacetate, PCMB	35,50,64,71,102,129
Calcium-activated neutral proteinase II	Cysteine (thiol)	—	Ca^{++} (μM), free sulfhydryl	EGTA, iodoacetate, PCMB	71,100
Neutral endopeptidase	Cysteine (thiol)	6.0 to 8.0	Free sulfhydryl	Iodoacetate, PCMB	3
Prolyl endopeptidase	Cysteine (thiol), serine	7.0 to 9.0	Free sulfhydryl	Iodoacetate, PCMB	104,119,161
Cation-sensitive neutral endopeptidase	Cysteine (thiol)	7.0 to 7.5	—	Na^+, K^+, PCMB	103,157,158
Plasminogen activator	Serine	7.0 to 9.0	—	DFP	76,136
Alkaline proteinase	—	5.0 to 9.0	—	Cu^{++}, Cr^{++}	118
Renin	Aspartic (carboxyl)	6.0 to 7.0	—	Pepstatin	43,44,55,63,106
Cathepsin B	Cysteine (thiol)	3.0 to 6.0	EDTA, free sulfhydryl	Iodoacetate, PCMB	139
Cathepsin D	Aspartic (carboxyl)	3.0 to 5.0	—	Pepstatin	2,25,154

roster of brain proteinases is still far from complete. Table 5-1 summarizes certain properties of brain proteinases that have been purified sufficiently to allow adequate identification. The term "proteinase" in this context refers to enzymes capable of hydrolyzing internal peptide bonds of proteins (13). Other endopeptidases active toward small peptides but that have not been shown to act on proteins are excluded.

Since many of the enzymes listed have been isolated in a highly purified state from brain, detailed studies of their interactions with cytoskeletal proteins are possible. Even at the level of purified enzymes acting on individual substrates, however, appropriate caution must be used in interpreting the interactions. Related enzymes with markedly different substrate specificities may partially copurify through extensive purification schemes. For example, renin and cathepsin D, two aspartyl proteinases that appear to serve very different functions, had proved troublesome to separate in earlier studies (51,54,106). The properties of some enzymes may be altered during purification to create the impression of multiple enzymes (122,134) or a novel enzyme (141,142). Recently, highly active forms of calcium-activated proteinase with increased calcium sensitivity and lower molecular weight were shown to be generated by autolysis of the enzyme purified from skeletal muscle (141,142). Sensitivity to leupeptin of calcium-activated neutral proteinase (CANP) has also been shown to change during purification (71). Furthermore, a dissociable subunit of CANP from skeletal muscle, present in variable amounts in different purified preparations, influences the activity of the major subunit (147). Endogenous activators of cathepsin D (61) may copurify until late purification steps. Similarly, endogenous inhibitors of many proteinases (14,147) including factors highly specific for CANPs (26,95,144,152) have been identified. The presence or absence of these additional modulating factors may potentially affect the specificity of the enzyme. Even apparently homogenous preparations of certain proteolytic enzymes may exhibit great complexity in isoenzyme composition. Most striking perhaps is cathepsin D, which is composed of up to eight isoenzymes in brain tissue (4). Two forms of calcium-activated neutral proteinase in CNS tissue similar in size (88,000- and 93,000-mol wt), have recently been resolved (71). This microheterogeneity and the demonstration of precursor species of certain proteolytic enzymes (40,97,115) raise the possibility that minor or short-lived enzymatic forms may have significant physiological roles *in vivo*.

Interpreting hydrolytic activity of a proteinase toward a given protein substrate also requires that the state of this substrate be considered. It is well known that the tertiary and quaternary structure of proteins greatly influences the rate of proteolysis (47,124) and that for many proteins the susceptibility to thermal denaturation is correlated with the degradative rate (18,88). Cathepsin D, the best-studied example (12), and CANP (138) may be more active toward denatured substrates than toward the corresponding native forms. Proteins that have undergone extensive or harsh preparative procedures prior to analysis may, therefore, have altered susceptibility to enzymes. Similarly, abnormally synthesized proteins may also be degraded more rapidly than their normal counterparts (47,48). In nonneural tissues, an ATP-dependent neutral-alkaline proteinase has been shown to be involved in the breakdown of abnormal proteins (41). These observations carry the important implication that in pathological states a proteolytic enzyme that does not normally act on a given protein may have a significant biological role when this protein is abnormal. Posttranslational modification (glycosylation, phosphorylation, etc.) (69) or interaction with a cofactor or substrate may also alter the susceptibility of enzymes and other proteins to proteolytic enzymes (47,48,124). This raises the possibility that certain proteins may only be substrates for a given proteinase under specific physiological conditions *in vivo*. When this substrate is then isolated for *in vitro* analysis, a resistant or highly susceptible form possibly could predominate and skew the results. Further suggested by these studies is the possibility that modification of the protein *in vitro*, as in radiolabeling procedures, may render the substrate more or less susceptible to attack by proteinases. Indeed, reductive alkylation of ϵ amino groups of lysine residues, a frequently used procedure for radiolabeling proteins, increases the susceptibility of the protein to hydrolysis (68). Finally, the conditions under which the reaction is performed may affect the state of the proteinase as well as the substrate. Cathepsin D, for example, may exhibit a bimodal activity curve toward denatured hemoglobin at different pH values between 3.0 and 4.5 (25, 27,106). Depending upon the pH of the reaction, rat brain cathepsin D cleaves different bonds in myelin basic protein (155).

Calcium-Activated Neutral Proteinases

Molecular Properties

Only several proteinases in brain tissue have been studied with respect to neurofilament proteolysis. Calcium-activated neutral proteinases have received by far the greatest attention. CANPs, purified from at least eight tissue sources (Table 5-2), exhibit various common features, including: (1) a pH optimum in the neutral to slightly alkaline range (7.0 to 8.7); (2) dependence on free sulfhydryl groups for activity; and (3) inhibition by the related peptide inhibitors leupeptin,

Table 5-2

Properties of Calcium-Activated Proteinases in Nonneural Tissues

Source (Species)	[Ca++] for Activity (mM)			Optimal pH	Molecular Weight		Enzyme Inhibitors			References
	Minimum	½ Maximum	Maximum		Subunit 1	Subunit 2	Sulfhydryl Reagents	Leupeptin	Serine Active Site	
Skeletal muscle (human) (monkey) (porcine) (rabbit)	0.1	0.8 to 2.8	1 to 5	6.9 to 8.0	80,000	30,000	+	+	–	7,30,31,62, 116,140,147
Skeletal muscle (chick)			1.8	7.5 to 7.8	80,000	–	+	+	–	66,67,138, 141
Smooth muscle (bovine)			4	8.7	115,000*	–			–	114
Cardiac muscle (canine)		0.8		7.0	135,000*	–			–	90,91
Liver (rat)		0.2	0.4	7.0 to 8.5	150,000*	–	+			32
Oviduct (chick)			0.5 to 1.0		113,000	–	+		–	149
Erythrocytes			3.3		80,000	–	+			28
Platelets			1.0	7.4	80,000	30,000	+	+		1,46
Skeletal muscle (porcine)	0.005	0.045		7.5 to 8.0	80,000	30,000	+	+	–	31
Skeletal muscle (chick)		0.040		7.5 to 7.8	76,000	–	+	+		77,141,142
Cardiac muscle (canine)		0.040		7.0	135,000*	–	+	+		91,90
Liver (rat)		0.007	0.025 to 0.050	7.0 to 8.5	150,000*	–	+			32

*Determined through gel-filtration studies only.

antipain, and E-64 but not by serine active-site inhibitors (PMSF, DFP) (reviews in 67, 131).

Recently, it has become apparent that calcim-stimulated proteolysis may reflect the activity of a family of proteolytic enzymes that can be distinguished by their molecular size and sensitivity to calcium (31,32,77,90,91). CANP in most tissues displays a molecular weight of 110,000 to 115,000 by gel filtration and consists of two subunits, 80,000-mol wt and 30,000-mol wt, when electrophoresed under denaturing conditions (Table 5-2). Curiously, although CANP from chick oviduct shows this pattern (149), its counterpart in chick skeletal muscle is isolated only as a monomeric 80,000-mol wt form (66,67,138,139). A different form of CANP with a molecular weight of 135,000 to 150,000 by gel filtration is present in liver (32) and cardiac tissue (90,91). In tissues containing CANP of either molecular size, two classes of CANP (CANP I and CANP II) are present and have calcium requirements in the millimolar (0.5 to 10 mM) and micromolar (7 to 70 μM) range, respectively (Table 5-2). Several observations suggest that these enzymes may be structurally related. In chick skeletal muscle, autolytic cleavage converts the 80,000-mol wt subunit of CANP I, the enzyme form active at millimolar calcium concentrations, into a 76,000-mol wt form, which requires micromolar calcium concentrations for maximal activity (141,142). Identity of this autocatalytic product with the CANP II that has been purified from different sources remains to be shown. In other non-neural tissues, the molecular weights of CANP I and CANP II are similar, although only gel filtration was used in two of three studies estimating the size of these forms (31,32,90,91). In skeletal muscle, where the forms differ by isoelectric point but not by molecular weight, antibodies raised against CANP I produced a single precipitant line with purified CANP II (31). In addition to these major forms, other calcium-dependent proteolytic activities have also been detected by ion-exchange chromatography, although their instability has prevented further analysis.

CANPs in nervous tissue (Table 5-3) share many properties with their counterparts in other tissues including (1) pH optimum (7.2 to 8.5), (2) sulfhydryl dependence, and (3) inhibitor profile. As in other tissues, at least two forms of CANP with differing calcium requirements are present in central nervous tissue (71,102). The major form (CANP I) requires greater than 1.0-mM calcium for maximal activity and is inactive below 0.1-mM calcium (Table 5-3). A second activity (CANP II) displays half-maximal activity at 20-μM calcium (71). The relative proportions of CANP A and CANP B, measured as maximal activity toward casein, are estimated to be 5: to 10:1, a low ratio compared with most other tissues (71). The molecular weight of rat

Table 5-3

Properties of Calcium-Activated Proteinases in Rat Nervous Tissue

Ca^{++} Requirement (mM)			Activity with Other Ions	Molecular Weight		Substrates	References
Minimum	½ Maximum	Maximum		Subunit 1	Subunit 2		
>0.1	—	2 to 5	Slight Sr^{++} Mn^{++}	93,000	—	Cyclic nucleotide-independent protein kinase	64
—	—	1	—	—	—	Casein, many proteins but no peptides	50
—	—	—	—	100,000	30,000	Neurofilament proteins	129
>0.1	—	—	<10% Sr^{++}, Mn^{++}	93,000	—	Broad specificity	71
—	0.020	—	<10% Sr^{++}, Mn^{++}	88,000	—	Broad specificity	71

brain CANP also differs from that in nonneural tissue. CANP I has been reported by separate laboratories to be a single subunit of 93,000-mol wt by gel filtration and SDS-PAGE (64,71). CANP II can be resolved from CANP I on SDS-PAGE as an 88,000-mol wt protein (71). Recently, isolation of CANP I by affinity chromatography using neurofilament protein as the ligand yielded a pure enzyme composed of 100,000-mol wt and 30,000-mol wt subunits (129).

Calcium-Stimulated Proteolysis of Neurofilaments in Invertebrates

Although Guroff initially identified CANP in rat brain (50), the first identification of its possible endogenous substrates in neurons was made through the use of isolated axoplasm from invertebrate giant axons. The earliest observations that exogenous calcium caused squid axoplasm to liquify (57,58) were followed by the demonstration of calcium-dependent protein hydrolytic activity in extruded squid axoplasm (105). In the giant axon of the marine worm *Myxicola infundibulum*, Gilbert and associates (46) showed that the calcium-dependent dispersal of isolated axoplasm was accompanied by the disintegration of neurofilaments into ill-defined flocculent clots. This process coincided with shifts in the molecular weight of neurofilament proteins on SDS-PAGE to smaller forms. Other axoplasmic proteins also gradually disappeared from the gels during this process, although NFPs decreased within 15 seconds of incubation at $37°C$. These events were interpreted as the action of a calcium-activated proteinase, since they were inhibited by sulfhydryl reagents and the protease inhibitors TPCK (L-1-tosylamide-2-phenylethylchloromethyl ketone) and TLCK (N-p-tosyl-L-lysine-chloromethyl-ketone). From recent studies by Pant and Gainer (108,110), it appears that more than one CANP activity requiring millimolar calcium concentrations (CANP A) may be present in squid axoplasm. Total extracted axoplasm contains calcium-dependent proteolytic activity that degrades most endogenous proteins but is especially active toward higher-molecular-weight species including the 200,000-mol wt neurofilament protein. Following high-speed centrifugation (100,000 g), however, the supernatant contains a CANP that selectively degrades 200,000-mol wt neurofilament protein while the pellet retains the relatively nonspecific CANP activity (108). Both activities require at least 0.5-mM calcium and optimally 4-mM calcium. The similarity in the action of this specific CANP to a form present in platelets (145) was demonstrated by the ability of the latter enzyme in a partially purified state to selectively degrade the 200,000-mol wt neurofilament protein.

The complete degradation of neurofilaments in invertebrate axoplasm appears to proceed through characteristic intermediate fragments. Early in the degradation of squid 200,000-mol wt neurofilament protein, a 100,000-mol wt fragment increases substantially in concentration and gradually proceeds through 47,000- to 55,000-mol wt forms before finally converting to acid-soluble peptides or amino acids (80,110). In *Myxicola*, the two major neurofilaments at 160,000- to 171,000-mol wt and 150,000- to 153,000-mol wt and the multiple minor components between 140,000-mol wt and 55,000-mol wt show close relationships by peptide analysis, which raise the possibility that the smaller forms are derived by proteolysis *in vivo* or during isolation (38,39). Indeed, it has been shown that ^{125}I-labeled neurofilament proteins from *Myxicola* break down into multiple components, including major products of 98,000- and 55,000- to 60,000-mol wt (80). Similar polypeptides are present as minor constituents of whole axoplasm prepared with EGTA, but these do not sediment with intact neurofilaments (80).

Calcium-Mediated Proteolysis of Neurofilaments in Vertebrates

Peripheral Nervous System

The calcium-induced disruption of neurofilaments in peripheral nerves has been extensively described from morphological and biochemical perspectives in Chapter 3 and will be treated only briefly here. Schlaepfer and associates have shown that, when excised peripheral nerves are incubated *in vitro* in the presence of millimolar calcium concentrations, the neurofilaments, microtubules, and other axonal organelles disappear and are replaced by an amorphous granular material (125,127). These events are accompanied by the loss of neurofilament proteins (200,000, 150,000, and 70,000-mol wt) identified by SDS-PAGE (125,128). These changes are minimal in the absence of exogenous calcium or in the presence of calcium chelators but are accelerated under conditions that favor rapid calcium influx into axons (e.g., freeze-thawing, calcium ionophore A23187, or Triton X-100) (125,127,128).

A soluble tissue factor that is likely responsible for these changes has been partially purified and shown to be a calcium-activated thiol-dependent enzyme or group of enzymes active between pH 6.0 and 8.0. Like the squid enzyme, maximal activity is achieved at millimolar concentrations. No changes in the specificity for individual ^{125}I]-labeled neurofilament proteins were noted when the enzyme was incubated with varying calcium concentrations (129,130).

In excised nerve segments and in partially purified enzyme preparations, the disappearance rate was highest for the 68,000-mol wt NFP, somewhat less for the 150,000-mol wt NFP, and considerably lower for the 200,000-mol wt form (128,129). When the incubating conditions favored a slow influx of calcium into axons of excised peripheral nerve, possible intermediate products of neurofilament breakdown at 70,000 to 73,000-mol wt and 53,000 to 57,000-mol wt were observed (128). Although it is especially active toward neurofilament proteins, CANP from peripheral nerves acts on tubulin, on several high- and low-molecular-weight proteins, as well as on casein used as a standard substrate.

Central Nervous Tissue

Calcium-mediated disruption of neurofilaments has recently been demonstrated in homogenates of spinal cord (9,60,126,130) and in brain extracts (83). The soluble tissue factor mediating these changes required more than 0.2-mM calcium for detectable activity (126) and exhibited properties similar to those of its apparent counterpart in peripheral nerve. In our laboratory, a different strategy enabled proteolysis to be studied specifically within axons of mouse retinal ganglion cell (RGC) neurons (98,99,100,101,102). By injecting mice intravitreally with radioactively labeled amino acids, the axoplasmic transport system of the RGC neuron can be exploited to segregate in the primary optic pathway a labeled protein population that resides specifically in axons. The average degradation rate of the labeled axonal proteins may then be determined *in vitro* within the excised but intact optic nerve and tract by measuring the release of labeled free amino acids and acid-soluble peptides from proteins under conditions that prevent amino acid reutilization. Alternatively, the loss of individual labeled axonal proteins may be determined by subjecting incubated and control optic pathway samples to SDS-PAGE and then quantitating the time-dependent loss of radioactivity within specific protein bands (100,102). By changing the pH and by adding ions or specific protease inhibitors during the incubation, the activity of individual protease activities may be selectively measured.

The techniques permitted two major proteolytic systems to be identified in RGC axons, each comprised of multiple enzymes (98). An "acidic" system optimally active when optic pathways are incubated at acidic pH (3.0 to 5.0), corresponds in its properties to the acid proteinase activity measured in brain homogenates (86,87). A "neutral" proteolytic system optimally active within the physiological pH range is also present in RGC axons. This system is stimulated more than threefold by adding exogenous calcium (1 to 4 mM) to the incubating medium (98,99). When maximally stimulated, neutral proteinase activity toward RGC axonal proteins is two to three times higher than total acidic proteolytic activity (98), a ratio similar to that in squid axoplasm (105).

At least three neutral proteinases have been shown to comprise the "neutral" proteolytic system in RGC axons (Table 5-4) (100). Among these, two are calcium-activated neutral proteinases. CANP A is maximally active at millimolar calcium concentrations (more than 1.0 mM) and within the pH range of 7.0 to 8.0. This enzyme activity is inhibited more than 90% by leupeptin (40 μg per milliliter), antipain (100 μg per milliliter), or sulfhydryl reagents (iodoacetate, parachloromercuribenzoate [PCMB], mersalyl, N-ethyl maleimide at 1 to 5 mM) and is partially inhibited by TLCK, zinc chloride, and copper chloride (1 mM). Pepstatin, TPCK, phenylmethylsulphonyl fluoride (PMSF), and diisopropylfluorophosphate (DFP) have minimal effect (less than 10% inhibition). In these respects, therefore, CANP A in RGC axons corresponds to the CANP of peripheral nerve. It is distinguishable from a second calcium-activated neutral proteinase in RGC axons (CANP B), which is leupeptin insensitive and PMSF sensitive (100, 102). (See below.) At millimolar concentrations of calcium, CANP A activity, defined as calcium-dependent activity inhibitable by leupeptin, represents at least 90% of the total calcium-dependent proteolytic activity in RGC axons (R. A. Nixon, unpublished observations).

In mouse RGC neurons, neutral proteinase(s) that are activated by millimolar calcium concentrations have broad specificity for axonal proteins. In the experiment illustrated in Figure 5-2, mice were injected intravitreally with [^{3}H]-proline, Seven days after the injection, the labeled optic pathways were removed and incubated at 37°C for 1 hour in HEPES medium, pH 7.4, containing either 4-mM calcium or 10-mM ethyleneglycol-bis(-aminoethyl ether)-N,N'-tetracetic acid (EGTA). One freeze-thaw cycle of the tissue prior to analysis facilitated the entry of exogenous calcium or of EGTA without altering the cellular specificity of the proteolytic measurements (99). Following incubation, proteins in each optic pathway were separated by SDS-PAGE and the radioactivity in consecutive 1-mm slices of each gel lane was determined. CANP activity is reflected by differences in the radioactive profile of optic pathways incubated with calcium and without calcium (plus EGTA). The results of this experiment demonstrate that virtually all RGC axonal proteins are susceptible to degradation by CANP(s) (102). High-molecular-weight proteins, however, appear to be relatively more susceptible. In fact, a highly significant correlation has been observed between the molecular size of the proteins (log Mr) and their rate of degradation by CANPs (Nixon and Froimowitz, in preparation). The correlation coefficient in this experiment was 0.787 $p < 0.001$. Similar results have been obtained by using shorter incubation intervals or fresh tissues.

Previous investigators have commented on the high susceptibility of neurofilament proteins to CANP (6,46,92,109,125,126,127,128, 130). In the foregoing study, the technique of radiolabeling axonally

Table 5-4

Properties of Neutral Proteinases in Axons of Mouse Retinal Ganglion Cells

Enzyme	pH Optimum[a]	EGTA[b]	Sulfhydryl Reagents[b]	Leupeptin[b]	PMSF[b]	Specificity[b]
Calcium-activated neutral proteinase A	6.0 to 9.0	+++	+++	+++	0(+)	Broad, especially neurofilament and other high-molecular-weight proteins
Calcium-activated neutral proteinase B	5.5 to 8.5	+++	+++	0(+)	++	Restricted, especially 145,000-mol wt neurofilament protein
Calcium-independent neutral proteinase	6.0 to 9.0	0	0	0	++	Broad, especially low-molecular-weight proteins

[a]Refers to the buffer pH range that is optimal for measuring enzyme activity in excised but intact optic pathway samples.
[b]Percent inhibition: 0, no effect; +, 1% to 10%; ++, 20% to 60%; +++, more than 60%.

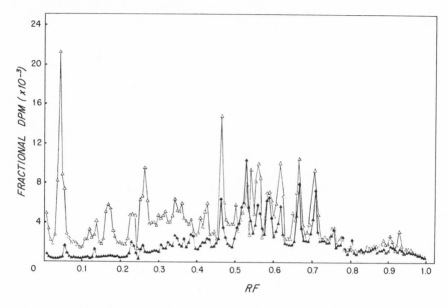

Figure 5-2. Specificity of calcium-activated neutral proteinase(s) toward proteins of mouse retinal ganglion cell axons. Excised optic pathways were incubated for 1 hour at 37°C in HEPES medium pH 7.4 (98) containing either 4-mM $CaCl_2$ (closed points) or 10-mM EGTA (open points). Following incubation, the tissues were subjected to SDS-PAGE. The radioactivity in consecutive 1-mm slices of each gel lane (the "fractional dpm") is expressed as the percentage of the total dpm in the optic pathway before incubation. The top of the gel is on the left side.

transported proteins provided a much larger population of high-molecular-weight proteins for analysis than was previously available in studies of Coomassie blue-stained proteins. Among this labeled group, at least several proteins unrelated to neurofilament proteins in structure and axoplasmic transport rate were degraded as rapidly or more rapidly than neurofilament proteins. Therefore, although CNS NFPs may be highly susceptible to CANPs, they cannot be considered a special protein class in this respect. These observations raise the possibility that the apparent selectivity of CANP A for neurofilament proteins in CNS neurons may be partly related to the relatively large size of NFPs or to another property shared by certain high-molecular-weight proteins.

In contrast to the broad substrate specificity of CANP(s) active at millimolar calcium concentrations, another calcium-dependent proteinase activity is present in mouse RGC axons that carries out the limited proteolysis of a single major neurofilament protein at endogenous calcium levels (100). This proteinase (CANP B) mediates the conversion of the 145,000-mol wt NFP to two lower forms at

143,000 and 140,000-mol wt. These three forms correspond in their two-dimensional electrophoretic patterns and peptide map profiles (Nixon, Brown, and Marotta, in preparation) to the 145,000-, 143,000-, and 1400,000-mol wt NFPs present in fresh mouse optic pathway (101) or in neurofilament-enriched preparations from whole mouse brain (21). Figure 5-3A depicts the 140,000- to 145,000-mol wt region of SDS gels from fresh optic pathways incubated for varying intervals in HEPES medium. The 145,000-mol wt NFP present in unincubated tissue is almost totally converted to 143,000- and 140,000-mol wt forms with similar isoelectric points (inset, Figures 5-3D and 5-3E) during the first 10 minutes of incubation (Figure 5-3A, lanes a-e). Changes in other axonal proteins on one- and two-dimensional gels were not detectable within this time interval (Figures 5-3B, 5-3C, and 5-3D). In studies of the radioactively labeled NFP species, the conversion from 145,000-mol wt to 143,000-mol wt and 140,000-mol wt forms was shown to be quantitative. The modification of the 145,000-mol wt NFP is activated by calcium since EGTA (10 mM) reversibly inhibits the process (Figure 5-3A, lane f). Thus, the enzyme may be highly active at levels of calcium (and other activating ions) that would normally exist within RGC axons. This conclusion is supported by the observation that a similar modification of the 145,000-mol wt NFP occurs in RGC axons during axoplasmic transport *in vivo* (101). (See below.)

The *in vitro* NFP modifications are partially blocked by the serine proteinase inhibitor PMSF, which suggests the involvement of a proteinase (100,101). This property differentiates the enzyme(s) from CANP A, which is unaffected by this inhibitor. In addition, CANP B is relatively insensitive to leupeptin and antipain, which completely inhibit CANP A. Like CANP A, CANP B is thiol dependent and optimally active when the incubations are conducted within the neutral pH range (102). As discussed later, these enzymes can be further distinguished by the predominant distal location of CANP B activity along RGC axons (101,102) and the relatively greater activity of CANP A in proximal axonal regions (R. A. Nixon, unpublished observations).

It is unlikely that the modification of 145,000-mol wt NFP represents an initial stage of degradation. Only the two forms at 143,000 and 140,000-mol wt are generated even after long incubation times. These forms are degraded *in vitro* by an enzyme activity with the same properties as CANP A (R. A. Nixon, unpublished data). In unincubated tissue, the 143,000- and 140,000-mol wt NFPs are major elements of axons that arise *in vivo* during axoplasmic transport and persist rather than convert to lower-molecular-weight fragments (101). The nature of these modifications *in vivo* and *in vitro*, therefore is

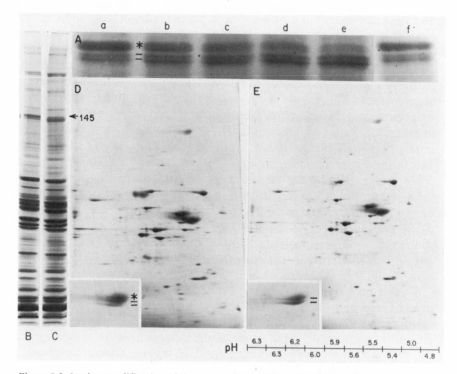

Figure 5-3. In vitro modification of the group of 140,000- to 145,000-mol wt neurofilament proteins in the primary optic pathway. Mice were injected intravitreally
with L-[³H] -proline. After 30 days, the excised but intact optic pathways from these
mice were incubated at 37°C in prewarmed HEPES medium (98), pH 7.4 (lacking
calcium). Following incubation, each optic pathway was analyzed by one- or two-dimensional SDS-polyacrylamide gel electrophoresis (polyacrylamide gradient in Figure
5-3A was 4% to 7%; in Figures 5-3B and 5-3C, 5% to 15%; and in Figure 5-3C and 5-3D,
8 to 18%) and the proteins were visualized with Coomassie brilliant blue. Figure 5-3A
displays only the region of the gel containing the 145,000- (asterisk), 143,000-
(upper bar), and 145,000- (lower bar) mol wt neurofilament proteins. (Lane a) Unincubated pathways; (lanes b, c, d, and e) optic pathways incubated 1, 2, 5, and 10 minutes
respectively; (lane f) a 10-minute incubation in the presence of 10-mM EGTA. Figure
5-3B: unincubated optic pathway. Figure 3C: optic pathway incubated for 10 minutes.
Figures 5-3D and 5-3E show respectively, optic pathways that were unincubated and
incubated for 30 minutes. In these figures, insets show the 145,000- to 140,000-
mol wt NFPs of each gel. (From reference 100. Used with permission.)

quite distinct from the rapid progressive cleavage of all neurofilament
species that is mediated by CANP A.

In summary, two calcium-activated thiol-dependent proteinases in
RGC axons are active toward neurofilament proteins. They can be
distinguished by their inhibitor profiles, requirements for calcium,
and specificity for axonal proteins. Analogies may be tentatively
drawn between these enzymes and the two partially purified CANPs

from rodent brain (71) that require millimolar and micromolar calcium levels, respectively, although the substrate specificity of the latter enzymes has not yet been established. The millimolar calcium requirement for CANP A creates an interesting problem in understanding how this enzyme functions in neurons where calcium levels are 10^{-7} to 10^{-8} M (8,19,82,112). Since calcium is not uniformly distributed in neurons (19,145), it is possible that CANP A is active only in specific neuronal sites. Similarly, CANP A may be activated only under certain physiological conditions where calcium levels transiently increase. It should also be appreciated, however, that under optimal conditions, CANP A can degrade 50% of the NFP in RGC axons in less than 30 minutes at 37°C (R. A. Nixon, unpublished observations). Therefore, even if less than 1% of its total activity were present at endogenous calcium concentrations, CANP A could contribute significantly to neurofilament metabolism. In any case, the broad specificity of CANP A points to a cellular function for this enzyme that is not restricted to neurofilaments. This contrasts with the exquisite specificity of CANP B for the 145,000-mol wt NFP at endogenous calcium concentrations that may indicate a more specialized role for this enzyme in neurofilament structure and function. As discussed below, other evidence indicates that this enzyme may be active toward the 145,000-mol wt neurofilament protein *in vivo* (101).

Calcium-Independent Neutral and Alkaline Proteinases

A variety of neutral and alkaline proteinases that do not require calcium have been isolated from nonneural tissues (reviews in 24,70, 112). Although relatively little attention has been paid to these enzymes in nervous tissue, several have been identified (See Table 5-1). Riekkinen and Rinne (118) partially purified an alkaline proteinase that optimally hydrolyzed hemoglobin and casein at pH 7.8 to 8.0. This enzyme differed in molecular weight, inhibitor profile, and pH optimum from the more active calcium-activated neutral proteinase. Unlike the latter enzyme, the alkaline proteinase was stimulated by EDTA and inhibited by divalent metal ions, particularly Ca^{++} and Cr^{++}. It was unaffected by DFP or trypsin inhibitors. Other neutral proteinases that are highly active toward small peptides also appear to degrade proteins. A thiol-dependent neutral endopeptidase that degrades hemoglobin has been purified from hypothalamus (3). It is leupeptin-insensitive but is inhibited by TPCK and slightly by PMSF. Prolyl endopeptidase, another thiol-dependent enzyme is markedly inhibited by calcium and by serine-active site inhibitors (104,119, 161). This enzyme hydrolyzes proline-containing peptide bonds not only of small peptides but also of proteins such as myelin basic protein. Plasminogen activator (76,136), a trypsin-like serine proteinase, has only one known macromolecular substrate—the zymogen plasminogen. Recently, this enzyme has been shown to be secreted at growth cones of cultured developing neurons (76). Finally, a calcium-independent neutral proteinase activity is present in mouse RGC axons. This enzyme or group of enzymes acts primarily on axonal proteins with molecular weights less than 60,000 (102).

Although the affinity of these enzymes for NFPs has not yet been reported, several observations suggest that calcium-independent proteinases may act on neurofilament proteins at neutral pH. In *Myxicola*, cleavage of small peptides from the neurofilament proteins in the absence of calcium is believed to involve proteinases active over a wide pH range (6.0 to 10.0) (39). Further characterization of this process, including specific evidence for a proteolytic enzymatic mechanism, has not yet been presented. Day (29) has described an enzyme activity in mammalian brain that generates lower-molecular-weight products from enriched neurofilament fractions at alkaline pH. When neurofilaments were solubilized from myelinated axons in 1-M sodium chloride at pH 9.5, a 58,000-mol wt protein appeared at the expense of the neurofilament proteins. Since PMSF prevented this process, a proteolytic mechanism was suspected. This process could be distinguished from a calcium-dependent proteolysis of NFPs that resulted in more extensive protein digestion.

Acidic Proteinases

Proteinases with pH optima in the acidic range are highly active in nervous tissue and display characteristics similar to their counterparts in nonneural tissues (reviews in 12,156). At least two major proteinases, cathepsin D and cathepsin B, contribute to this acidic proteolytic activity. In nonneural tissues, these enzymes and other proteinases with acidic pH optima have been demonstrated by histochemical and immunocytochemical techniques to be located within lysosomes (15, 93,132). Evidence from studies of nervous tissues, although still incomplete, also points to a parallel distribution of cathepsin D and lysosomes within neurons (156). Whitaker (156) has demonstrated by immunochemical techniques that cathepsin D is enriched in neurons compared to other neural cells and varies markedly in content within different neuron types. Immunoreactivity and enzyme activity are predominantly localized within neuronal perikarya (56,132, 156) but are detectable in neuronal processes as well (156). Acidic proteolytic activity in the axons of mouse retinal ganglion cell neurons assumes a proximodistal gradient with highest activity in proximal axonal sites (98). Specific measurement of cathepsin D indicates a similar gradient with two- to threefold higher activity proximally than distally (R. A. Nixon, unpublished observations). These observations accord with the apparent distribution of lysosomes within neurons. Although lysosomes have been observed in synaptic terminals, they are present only infrequently (20,81).

Cathepsin D has been purified from the central nervous system of many mammalian species (review in 87; also 3,25,154). The major molecular form has a molecular weight of 40,000 to 43,000, although variable amounts of a 30,000- to 35,000-mol wt and 14,000- to 16,000-mol wt protein are also present in most species. It is unclear whether these smaller proteins, which may represent either subunits or proteolytic fragments, are catalytically active. At least five isoenzymes of brain cathepsin D have been identified, each displaying similar behavior with respect to enzyme inhibitors (4). The existence of proforms (115) and glycosylated species (40) of cathepsin D in nonneural tissues is a precedent for suspecting even greater complexity of the brain enzyme than has been reported. These molecular properties of cathepsin D, its distribution within the neuron, and differential concentration among neuronal types suggest that this enzyme serves multiple roles in normal or pathological processes.

Purified cathepsin D is highly active toward neurofilament proteins. Cathepsin D from postmortem human brain was purified to homogeneity by ion-exchange and affinity chromatography (25). A single peak of proteolytic activity toward hemoglobin was detected by gel filtration. Under denaturing conditions, the enzyme appeared on polyacrylamide gels as a 30,000-mol wt protein and a 15,000-mol wt doublet in equimolar concentrations. When cathepsin D was incubated with neurofilaments purified by axon flotation from human white matter (22), the major 200,000-mol wt, 160,000-mol wt, and 70,000-mol wt neurofilament proteins were appreciably degraded within 1 hour of incubation (Figure 5-4). Pepstatin, a specific inhibitor of cathepsin D, completely blocked NFP degradation (not shown). No striking differences were noted in the degradation rates of major NFPs by cathepsin D, although quantitative measurements have not been done. Intermediate products of degradation at 60,000 to 160,000 mol wt were observed, some of which seemed to correspond to minor bands present in the unincubated preparation. In addition to NFPs, three high-molecular-weight proteins and certain smaller ones were rapidly degraded. Not all proteins in the neurofilament preparation were equally susceptible. For example, a 49,000-mol wt protein, previously identified in neurofilament preparations as glial fibrillary acidic protein (22), was only slightly degraded by cathepsin D. These experiments, therefore, demonstrated that cathepsin D actively degrades neurofilament proteins. Whether or not this interaction occurs

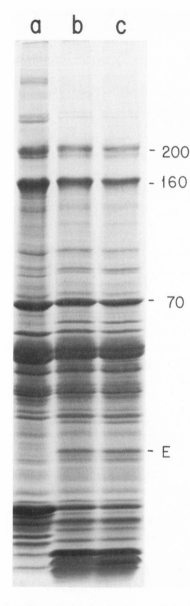

Figure 5-4. The activity of cathepsin D purified to homogeneity from postmortem human cortex toward enriched fractions of neurofilament proteins prepared by axonal flotation from human white matter. NFPs (7 mg) without added enzyme were analyzed after 2 hours of incubation at 37°C (lane a), in 0.15 ml of 10-mM ammonium acetate buffer at pH 3.2. NFPs were incubated with cathepsin D (3 mg of hemoglobin-degrading activity per hour) at 37°C for 1 hour (lane b) or 2 hours (lane c) in the same buffer. The 30,000-mol wt form of cathepsin D is indicated by the letter E.

under physiological conditions when the proteinase is compartmentalized in lysosomes is under investigation.

Cathepsin B has been isolated from several tissue sources (11,144), including brain (139). This thiol proteinase in brain is leupeptin-sensitive and partially inhibited by PMSF but unaffected by pepstatin. Its molecular weight (27,000), pH optimum (6.5) and immunoreactivity

(73) are similar to those of cathepsin B from other tissues. Brain cathepsin B has broad specificity for low-molecular-weight proteins and peptides, although some proteins appear to be resistant. Cathepsin B from other sources has also been shown to act on high-molecular-weight proteins such as myosin (205,000) (15). Its activity toward neurofilament proteins, however, remains to be tested.

Proteolysis of Neurofilament Proteins *In Vivo*

Posttranslational Modification of Neurofilament Proteins by Limited Proteolysis *in Vivo*

The presence of multiple proteinases along RGC axons raises the possibility that proteolysis within axons is an active process *in vivo*. We earlier hypothesized that, if proteolytic modification occurs as proteins are transported along axons *in vivo*, it may be evident as alterations of protein composition along the length of these axons (98,101). Recent observations have supported this possibility. When consecutive 1.1-mm segments of mouse primary optic pathway were analyzed by one- and two-dimensional SDS-PAGE, certain proteins exhibited a gradient of change—an increase or decrease in relative concentration—along the optic pathway (21,101). Among the most prominent changes were those in tubulins (21) and in the multiple forms of the 140,000- to 145,000-mol wt neurofilament protein (101). Figure 5-5, depicting the 140,000- to 145,000-mol wt neurofilament protein region of SDS gels, demonstrates that the two major NFP species (145,000 mol wt and 140,000 mol wt) are present along the entire length of the mouse primary optic pathway. A third NFP, one of the major forms in purified neurofilament preparations (21), is absent proximally but becomes increasingly prominent in distal segments of the optic pathway. Inspection of this figure also suggests decreased proportions of 145,000-mol wt NFP and possibly increased 140,000-mol wt NFP progressively along the optic pathway. Figure 5-6 summarizes further studies on the fate of newly synthesized neurofilament proteins within RGC axons. Following intravitreal injection of [3H]-proline, radioactively labeled NFPs in the 140,000- to 145,000-mol wt range entered proximal RGC axons as two major species corresponding to the 145,000-mol wt and 140,000-mol wt NFPs observed on stained gels (e.g., 6 to 15 days postinjection) (Figures 5-6A and 5-6B). When transported NFPs reached distal axonal regions (30 days postinjection or longer), a 143,000-mol wt protein began to accumulate; it was similar in its isoelectric point and peptide map pattern to the 143,000-mol wt and 140,000-mol wt NFP species (Figures 5-6C and 5-6D). Recent quantitative studies also demonstrate

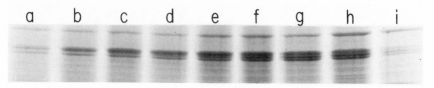

Figure 5-5. The group of 140,000- to 145,000-mol wt neurofilament proteins in consecutive segments of the optic pathway. Optic pathway was dissected and cut into 1.1-mm segments, and the proteins of each segment were analyzed by SDS gel electrophoresis. Segment 1 (lane a) was adjacent ot the scleral surface of the eye, segment 5 (lane e) includes optic chiasm, and segment 9 (lane i) was overlying the lateral geniculate body. Only the 140,000- to 145,000-mol wt neurofilament proteins are shown. These proteins are separated into at least two bands (140,000, 145,000 mol wt) in proximal segments (lanes a-c) and at least three bands (140,000, 143,000, 145,000 mol wt) in distal segments (lanes d to i).

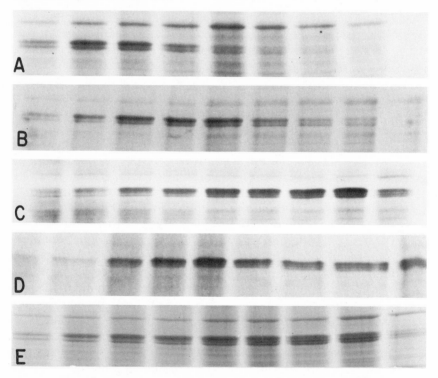

Figure 5-6. Radiolabeled 140,000- to 145,000-mol wt neurofilament proteins in consecutive segments of the mouse optic pathway (101). Proteins were labeled by intravitreal injection of L-[^{3}H] -proline. After various postinjection intervals, the optic pathway was excised and cut into 1.1-mm segments. Each segment was applied to a separate lane of a SDS gel. After electrophoresis, radioactive proteins were visualized by fluorography. The 140,000- to 145,000-mol wt NFP region of each gel is shown (A) 6 days, (B) 15 days, (C) 39 days, and (D) 60 days after injection of radioactive proline. The amounts of radioactivity loaded per gel lane ranged from 100,000 to 700,000 dpm. Panel E shows the pattern of the stained gel. (From reference 101. Reproduced from *The Journal of Cell Biology*, 1982, Vol. 94, pp. 150-158, by copyright permission of The Rockefeller University Press.)

that the 145,000-mol wt NFP decreases relative to the 143,000-mol wt and 140,000-mol wt species as these proteins move down the RGC axons (R. A. Nixon, unpublished results). These observations indicate that posttranslational modification of the 140,000- to 145,000-mol wt neurofilament proteins occurs during axoplasmic transport. This process leads to gradual loss of the 145,000-mol wt NFP and generation of the 143,000-mol wt and 140,000-mol wt NFP species along RGC axons.

The *in vivo* alterations of the 140,000- to 145,000-mol wt NFPs bear a striking resemblance to the ones mediated *in vitro* by CANP B (100). As previously discussed, this proteinase cleaves the 145,000-mol wt NFP and thereby generates 143,000-mol wt and 140,000-mol wt NFP species. These products exhibit peptide map patterns and isoelectric points on two-dimensional gels that are identical to the 143,000-mol wt and 140,000-mol wt proteins generated *in vivo* (Nixon, Brown, and Marotta, in preparation). Similarities between the *in vivo* and *in vitro* process have also been noted in the distribution of this enzymatic activity along the RGC axons. In both cases, activity appears to be considerably higher in distal RGC axons than at proximal sites (101). Finally, it is worth pointing out that the optimal conditions for promoting the NFP modifications *in vitro*, including the calcium requirement, are those that most closely resemble the normal *in vivo* state.

The foregoing observations suggest that these modifications of 140,000- to 145,000-mol wt NFP in RGC axons are mediated by a related enzymatic mechanism *in vivo* and *in vitro*. Implied, therefore, is the possibility that limited proteolysis is an active process in RGC axons *in vivo* and plays a special role in the normal functioning of neurofilaments. In view of the striking proximodistal variations in proteinase activities (98,101) and in cytoskeletal proteins other than NFPs (21,101) along RGC axons, we have proposed that CNS axons may be regionally specialized with respect to structure and function (discussed further in 101). As the role of neurofilaments becomes defined, it will be interesting to consider how the function of the axoplasmic matrix is influenced by this regional specialization.

Neurofilament Protein Degradation *in Vivo*

Under steady-state conditions, the prodigious synthesis of neurofilament proteins (16,159) must be balanced by an equally active catabolic mechanism. Weiss and Hiscoe (153) initially suggested that this degradative process may occur as cytoskeletal proteins are transported along the axon. According to this hypothesis, surplus cytoskeletal proteins would be eliminated before they reached axon terminals. Although proteolysis may indeed be active within the axon (see below), evidence has shown that a fraction of most axonally transported proteins reaches axonal terminals and appears to be metabolized there (reviews in 33,34,49). Recent observations indirectly suggest that a major portion of neurofilament proteins may also be actively degraded in synaptic terminals. Neurofilaments are far less frequent, and often absent, in synaptic terminals (107,113). In squid synaptosomes, the major neurofilament protein (200,000-mol wt) is absent (109). Moreover, long after intravitreal isotope administration when NFPs should have reached nerve terminals of RGC neurons, Lasek and Black (78) found only low levels of radioactivity associated with transported NFPs within the superior colliculus, a brain region enriched in axon terminals. This observation has been interpreted as evidence for the metabolism of NFPs entering the presynaptic area. Recently, Karlsson, and associates demonstrated *in vitro* degradation of rapid- and slow-phase proteins in synapses and preterminal axons that were preserved within slices of the superior colliculus (120). Proteolysis was greatly stimulated by millimolar concentrations of calcium. In this regard, it has been suggested (78,79) that changes in calcium transport into the specialized synaptic terminal region during physiological processes are responsible for raising synaptic calcium to levels necessary to sustain NFP degradation by CANPs. As noted earlier, the high potential activity of CANP A implies that suboptimal activation may be sufficient to support this degradative process.

The foregoing evidence is compatible with a hypothesis advanced by Lasek and associates that neurofilaments are constituents of a continually moving column of axoplasmic matrix that grows from the perikaryal end and is disassembled and degraded only at the synaptic terminal (78,79). This model of neurofilament metabolism has several implications. First, if NFPs are pulse labeled by injecting radiolabeled amino acids in the region of neuronal cell bodies, they should be conveyed along the axon as a coherent wave that does not broaden as transport proceeds. These investigators have demonstrated that in the phrenic nerve of the guinea pig, the transport of NFP between 32 and 64 days after injection approximates the behavior predicted by this hypothesis (78). Similarly, in rat sciatic nerve only a modest broadening of the wave is observed between 6 and 33 days postinjection, when it is still at relatively proximal sites along the nerve (59). At a longer survival time (60 days), however, appreciable broadening of the NFP wave with fragmentation into multiple peaks is apparent from their data (59). This phenomenon was explained by postulating that the rate of movement of the slow component varied in motor axons of differing caliber in the sciatic nerve. Changes in the form and coherence of the slow wave during progression along axons of sensory nerves (rat and cat dorsal roots) and of dorsal columns have also been noted by Stromska and Ochs (137), who interpreted this to indicate local deposition of transported proteins. In shorter central axons, this kind of analysis is more complicated, although it may be noted that, even at survival times up to 100 days postinjection in the guinea pig optic system, a small but significant amount of NFP is still present in the most proximal 3 mm of optic nerve (16,17). These neurofilament proteins appear to be deposited within axons or, at least, transiently dissociated from the major NFP wave. At progressively later times, the trailing edge of the NFP wave did not move coherently or at a rate as fast as observed for the leading edge. Apart from these observations, it remains to be demonstrated that the *main* wave of radiolabeled NFPs eventually disappears completely from the optic nerve and tract as predicted by the proposed model. Preliminary studies in mouse RGC neurons corroborate the finding that a certain fraction of NFP persists along axons long after the NFP wave should have completely entered nerve terminals (R. A. Nixon, unpublished observations). It also seems unlikely from our experiments that continued perikaryal synthesis of NFPs from reutilized labeled amino acids could account for the quantity of NFP that persists in proximal axons after long postinjection intervals. These findings suggest, therefore, that some of the neurofilament protein dissociates from the main wave of transported NFPs and is transiently or permanently deposited along certain axons.

The model of NFP metabolism proposed by Lasek and associates also implies that proteolysis of NFPs is absent along the axon during transport. This concept is supported by the observation that the total amount of radioactivity contained in the slow component (neurofilament proteins, tubulin, and associated proteins) does not change between 32 and 64 days postinjection in the phrenic nerve (78). However, due to the variation in the replicate determinations at each time-point (e.g., at 32 days, the S.E.M. was 27% of the mean value, $n = 5$), only a large decline in radioactivity would have been detectable in this experiment. In other studies, substantial decreases (more than 60%) in the radioactivity of the slow component can be discerned in rat sciatic nerve between 20 and 85 days (59), although it is possible that part of the decrease reflects transport of proteins into nerve branches that were not analyzed. Observations in guinea pig retinal ganglion cell axons are also consistent with significant radioactivity decline in slow-phase components during axoplasmic transport (16, 17).

If a fraction of the NFP pool is deposited within the axon, it may either be metabolized there or may eventually reassociate with the moving wave of neurofilament proteins. Neither possibility has been directly addressed experimentally, although indirect evidence suggests that protein breakdown may occur to some extent with axons *in vivo*. For example, proximodistal gradients in the proportions of a variety of proteins are present along RGC axons, and several proteins have been shown to decrease markedly during axoplasmic transport (22, 101; R. A. Nixon, unpublished results). Willard and associates have reported a short half-life (4 to 6 hours) for a protein that appears to be deposited along regenerating toad RGC axons (133). These investigators have also reported labeling of glial proteins in the optic tract at long survival intervals after intravitreal injection of isotopic amino acid (159). After ruling out that glial protein labeling was due to incorporation of blood-borne isotope, they concluded that reutilization of amino acids released by proteolysis in axons may be responsible. In this regard, optic nerve glia reutilize more than half of the radiolabeled amino acids released by the degradation of RGC axonal proteins *in vitro* unless protein-synthesis inhibitors are present (98).

The foregoing observations in this section are compatible with alternative mechanisms of neurofilament transport and metabolism. Varying amounts of NFP deposition along axons may be reconciled with an axoplasmic transport mechanism in which the transported neurofilaments or their constituent proteins may exchange with or replace metabolized components of a stationary axoplasmic matrix. This general mechanism and others are discussed in Chapter 4. Neurofilament proteins might be conveyed coordinately as separate subunits or as preassembled neurofilaments that are either unassociated with or may be freely dissociated from other transported components of the axoplasm. In this regard, the transport of neurofilament proteins may take place without the normal coordinate transport of tubulin in rat dorsal root ganglia cells treated with colchicine (74). Furthermore, in mouse retinal ganglion cells, neurofilament proteins are transported somewhat more rapidly than tubulin (R. A. Nixon, unpublished results).

Depending upon the metabolic needs of the axon, neurofilaments or neurofilament proteins would dissociate from the main transport wave and become incorporated along the axon. The extent of neurofilament protein deposition would be related to the turnover rate of these proteins in the axonal compartment. The hypothesis that neurofilaments are degraded partly within axons may account for a number of disparate observations in axoplasmic transport studies. In axons with a low NFP-turnover rate, for example, relatively little radioactivity from a wave of radiolabeled neurofilament proteins may be deposited along the axons. The dimensions of the wave would, therefore, be relatively conserved during transport. Varying degrees of spreading and diminution of the major wave of transported protein would reflect higher rates of turnover of neurofilaments within the axon. Neurofilament proteins that had not been incorporated along the axon would be degraded rapidly in the synaptic terminal to prevent accumulation. In normal axons, unincorporated NFPs are likely to be the major fraction of the total transported neurofilament. The need for such an apparent surplus of neurofilament may be appreciated from the following considerations. For a given transported protein, the *maximum* fraction of the initial wave reaching the synapse is described by the expression

$$\frac{1}{2^{\ell/rt}}$$

where ℓ is the axonal length, r is the rate of axoplasmic transport, and t is the half-life of the protein within the axon. The quantity ℓ/rt denotes the number of protein half-lives expended during the period of transit from the cell body to the synapse (ℓ/r). The level calculated from this expression would be further reduced by an amount equal to the quantity of protein deposited in axons. It may be seen from this general expression that the fraction of the initial protein wave reaching the synaptic terminal decreases exponentially as the half-life of the protein is reduced. A similar exponential decrease in this fraction occurs as the rate of axoplasmic transport declines. It also follows from this expression that, if protein turnover or axoplasmic transport were altered, the protein fraction reaching terminals of long axons would be more profoundly affected than that in shorter ones. The impact of a change in protein turnover rate may be illustrated by considering an axon of 200-mm length in which neurofilament proteins were transported at a rate of 1 mm per day. Choosing a hypothetical half-life for neurofilament proteins of 50 days, the maximum fraction of neurofilament protein reaching the synapse would be 12.5% of the initial wave. If the turnover rate doubled, however, this surplus protein fraction would be reduced to 0.4%, a 32-fold decrease. Neurons with longer neurites may be expected then to be especially vulnerable to pathological events involving increased proteolysis or impaired axoplasmic transport unless a substantial excess of the affected proteins was normally present.

According to the foregoing hypothesis, therefore, the fraction of neurofilament protein reaching the synapse may be viewed as the functional reserve of the axon that is available for axonal repair whether related to normal axon function or to metabolic and toxic insults. Active proteolysis in the synaptic terminals would clearly be needed to prevent accumulation of this reserve fraction. This hypothesis further suggests that neurofilament proteins that are deposited and presumably metabolized in axons, however small a percentage of the total transported wave, may represent the fraction involved in the normal maintenance of neurofilament organization within axons.

In summary, it seems reasonable to conclude that proteolysis of certain proteins is active in axons and synaptic terminals *in vivo*, although the relative activity of this process at these sites remains to be determined. Limited proteolytic modification of the 145,000-mol wt neurofilament protein is also likely within axons *in vivo*. Evidence for the degradation of neurofilament proteins at either neuronal site, however, is indirect. Several proteinases active toward NFPs are present in both locations, and it seems possible that proteolytic processing, including complete degradation, may occur to some extent within axons and synaptic terminals. Whether NFPs are also proteolytically processed in the cell body before they become committed to axoplasmic transport remains an important unanswered question.

Conclusions

Interest in the proteolysis of neurofilaments has accelerated rapidly in the past few years. Although the investigations are at an early stage, the impression emerges that proteolysis may be involved in several facets of neurofilament organization and function. At least two proteinases are able to initiate NFP breakdown to fragments that are ultimately degraded to amino acids. Particularly intriguing is the enormous potential activity of calcium-activated neutral proteinases and their apparent affinity for neurofilament proteins. At the present time, one can only speculate about how this catabolic capability is harnessed. Many specific questions remain about the high calcium requirement of these enzymes, the distribution and properties of microheterogeneous CANPs, and the role of specific endogenous inhibitors. The participation of the lysosomal system in neurofilament metabolism is still relatively unexplored even though, in theory, no obstacles would appear to militate against a significant involvement. The activity of cathepsin D and possibly other lysosomal proteinases toward neurofilament proteins warrants further investigation under conditions that preserve subcellular organization.

In addition to the complete breakdown of neurofilament proteins, mechanisms exist for the limited proteolysis of neurofilament proteins in CNS axons. A second calcium-activated neutral proteinase, operating at endogenous calcium levels, specifically modifies the group of NFPs of 140,000 to 145,000-mol wt, thereby creating microheterogeneity and varied proportions of these NFP forms along retinal ganglion cell axons. This enzyme appears to be active *in vivo* within RGC axons. The physiological significance of this process is unclear, although the possibility has been raised that limited proteolysis modifies the functioning of neurofilaments in different parts of the neuron.

REFERENCES

1. Airhart, J., A. Vidrich, and E. A. Khairallah. 1974. Compartmentation of free amino acids for protein synthesis in rat liver. *Biochem. J. 140*:539-548.

2. Akopyan, T. N., A. A. Arutunyan, A. Lajtha, and A. A. Galoyan. 1978. Acid proteinase of hypothalamus: Purification, some properties, and action on somatostatin and substance P. *Neurochem. Res. 3*:89-99.

3. Akopyan, T. N., A. A. Arutunyan, A. I. Oganisyan, A. Lajtha, and A. A. Galoyan. 1979. Breakdown of hypothalamic peptides by hypothalamic neutral endopeptidase. *J. Neurochem. 32*:629-631.

4. Akopyan, T. N., N. A. Barchudaryan, L. V. Karabashyan, A. A. Arutunyan, A. Lajtha, and A. A. Galoyan. 1979. Hypothalamic cathepsin D: Assay and isoenzyme composition. *J. Neurosci. Res. 4*:365-370.

5. Ames, A., III, J. M. Parks, and F. B. Nesbett. 1980. Protein turnover in retina. *J. Neurochem. 35*:131-142.

6. Anderton, B. H., C. W. Bell, B. J. Newby, and D. S. Gilbert. 1976. Neurofilaments. *Biochem. Soc. Trans., 53rd Meeting, London 4*:544-547.

7. Azanza, J.-L., J. Raymond, J.-M. Robin, P. Cottin, and A. Ducastaing. 1979. Purification and some physico-chemical and enzymic properties of a calcium ion-activated neutral proteinase from rabbit skeletal muscle. *Biochem. J. 183*:339-347.

8. Baker, P. F. 1972. Transport and metabolism of calcium ions in nerve. *Prog. Biophys. Mol. Biol. 24*:177-223.

9. Banik, N., E. Hogan, J. M. Powers, and L. Whetstine. 1982. Degradation of neurofilaments in spinal cord injury. *Trans. Am. Soc. Neurochem. 13*:256 (abs.).

10. Barondes, S. H. 1974. Synaptic macromolecules: Identification and metabolism. *Ann. Rev. Biochem. 43*:147-168.

11. Barrett, A. J. 1973. Human cathepsin B1: Purification and some properties of the enzyme. *Biochem. J. 131*:809-822.

12. Barrett, A. J. 1977. Cathepsin D and other carboxyl proteinases. *In*: Proteinases in Mammalian Cells and Tissues. A. J. Barrett, editor. North Holland, Biomedical Press, Amsterdam, pp. 209-248.

13. Barrett, A. J. 1980. Introduction: The classification of proteinases. *In*: Protein Degredation in Health and Disease. Ciba Foundation Symposium 75, Excerpta Medica, pp. 1-13.

14. Baugh, R. J., and H. P. Schnebli. 1980. Role and potential therapeutic value of proteinase inhibitors in tissue destruction. *In*: Proteinases and Tumor Invasion. P. Strauli, A. J. Barrett, and A. Baici, editors. Raven Press, New York, pp. 158-180.

15. Bird, J. W. C., A. M. Spanier, and W. N. Schwartz. 1978. Cathepsins B and D: Proteolytic activity and ultrastructural localization in skeletal muscle. *In*: Protein Turnover and Lysosome Function. H. L. Segal and D. J. Doyle, editors. New York, Academic Press, pp. 589-604.

16. Black, M. M., and R. J. Lasek. 1979. Axonal transport of actin: Slow component b is the principal source of actin for the axon. *Brain Res. 171*:401-413.

17. Black, M. M., and R. J. Lasek. 1980. Slow components of axonal transport: Two cytoskeletal networks. *J. Cell Biol. 86*:616-623.

18. Bond, J. S. 1975. Relationship between inactivation of an enzyme by acid or lysosomal extracts and its *in vitro* degradation rate. *Fed. Proc. 34*:651 (abs.).

19. Brinley, F. J., Jr. 1980. Regulation of intracellular calcium in squid axons. *Fed. Proc.* 39:2778-2782.

20. Broadwell, R. D., and M. W. Brightman. 1979. Cytochemistry of undamaged neurons transporting exogenous protein *in vivo. J. Comp. Neurol. 185*:31-74.

21. Brown, B. A., R. A. Nixon, and C. A. Marotta. 1982. Post-translational processing of alpha-tubulin during axoplasmic transport in CNS axons. *J. Cell Biol. 94*:159-164.

22. Brown, B. A., R. A. Nixon, P. Strocchi, and C. A. Marotta. 1981. Characterization and comparison of neurofilament proteins from rat and mouse CNS. *J. Neurochem. 36*:143-153.

23. Cicero, T. J., and B. W. Moore. 1970. Turnover of the brain specific protein, S-100. *Science 169*:1333-1334.

24. Clark, M. G., C. J. Beinlich, E. E. McKee, J. A. Lins, and H. E. Morgan. 1980. Relationship between alkaline proteolytic activity and protein degradation in rat heart. *Fed. Proc. 39*:26-30.

25. Compaine, A., and R. A. Nixon. 1981. Purification and some properties of cathepsin D from human brain. *Trans. Am. Soc. Neurochem. 12*:302 (abs.).

26. Cottin, P., P. L. Vidalenc, and A. Ducastaing. 1981. Ca^{2+}-dependent association between a Ca^{2+}-activated neutral proteinase (CaANP) and its specific inhibitor. *FEBS Lett. 136*:221-224.

27. Cunningham, M., and J. Tang. 1976. Purification and properties of cathepsin D from porcine spleen. *J. Biol. Chem. 251*:4528-4536.

28. Dahlqvist-Edberg, U., and P. Ekman. 1981. Purification of a Ca^{2+}-activated protease from rat erythrocytes and its possible effect on pyruvate kinase *in vivo. Biochim. Biophys. Acta 660*:96-101.

29. Day, W. A. 1980. CNS neurofilament proteolysis and the 58,000-dalton fragment. *J. Ultrastruct. Res. 70*:1-7.

30. Dayton, W. R., and J. V. Schollmeyer. 1981. Immunocytochemical localization of a calcium-activated protease in skeletal muscle cells. *Exp. Cell Res. 136*:423-433.

31. Dayton, W. R., J. V. Schollmeyer, R. A. Lepley, and L. R. Cortex. 1981. A calcium-activated protease possibly involved in myofibrillar protein turnover. *Biochim. Biophys. Acta 659*:48-61.

32. De Martino, G. N. 1981. Calcium-dependent proteolytic activity in rat liver: Identification of two proteases with different calcium requirements. *Arch. Biochem. Biophys. 211*:253-257.

33. Droz, B. 1973. Renewal of synaptic proteins. *Brain Res. 62*:383-394.

34. Droz, B. 1975. Synthetic machinery and axoplasmic transport: Maintenance of neuronal connectivity. *In*: The Nervous System. D. B. Tower, editor-in-chief. Vol. 1. The Basic Neurosciences. Raven Press, New York, pp. 111-127.

35. Drummond, G. I., and L. Duncan. 1968. On the mechanism of activation of phosphorylase *b* kinase by calcium. *J. Biol. Chem. 243*:5532-5538.

36. Dunlop, D. S. 1978. Measuring protein synthesis and degradation rates in CNS tissue. *In*: Research Methods in Neurochemistry. N. Marks and R. Rodnight, editors. Vol. 4. Plenum Press, New York, pp. 91-141.

37. Dunlop, D. S., W. van Elden, and A. Lajtha. 1978. Protein degradation rates in regions of the central nervous system *in vivo* during development. *Biochem. J. 170*:637-642.

38. Eagles, P. A. M., D. S. Gilbert, and A. Maggs. 1979. Neurofilament structure and enzymic modification. *Trans. Biochem. Soc. 7*:484-487.

39. Eagles, P. A. M., D. S. Gilbert, and A. Maggs. 1981. The polypeptide composition of axoplasm and of neurofilaments from the marine worm *Myxicola infundibulum. Biochem. J. 199*:89-100.

40. Erickson, A. H., and G. Blobel. 1979. Early events in the biosynthesis of the lysosomal enzyme cathepsin D. *J. Biol. Chem. 254*:11771-11774.

41. Etlinger, J., and A. Goldberg. 1977. A soluble ATP-dependent proteolytic system responsible for the degradation of abnormal proteins in reticulocytes. *Proc. Natl. Acad. Sci. USA 74*:54-58.

42. Fern, E. B., and P. J. Garlick. 1974. The specific radioactivity of the tissue free amino acid pool as a basis for measuring the rate of protein synthesis in the rat *in vivo*. *Biochem. J. 142*:413-419.

43. Fishman, M. C., E. A. Zimmerman, and E. E. Slater. 1981. Renin and angiotensin: The complete system within the neuroblastoma x glioma cell. *Science 214*:921-923.

44. Fuxe, K., D. Ganten, T. Hokfelt, V. Locatell, K. Poulsen, G. Stock, E. Rix, and R. Taugner. 1980. Renin-like immunocytochemical activity in the rat and mouse brain. *Neurosci. Lett. 18*:245-250.

45. Garlick, P. J., and I. Marshall. 1972. A technique for measuring brain protein synthesis. *J. Neurochem. 19*:577-583.

46. Gilbert, D., B. J. Newby, and B. H. Anderton. 1975. Neurofilament disguise, destruction and discipline. *Nature 256*:586-589.

47. Goldberg, A. L., and J. F. Dice. 1974. Intracellular protein degradation in mammalian and bacterial cells. Part 1. *Ann. Rev. Biochem. 43*:835-870.

48. Goldberg, A. L., and A. C. St. John. 1976. Intracellular protein degradation in mammalian and bacterial cells: Part 2. *Ann. Rev. Biochem. 45*:747-803.

49. Grafstein, B., and D. S. Forman. 1980. Intracellular transport in neurons. *Physiol. Rev. 60*:1167-1283.

50. Guroff, G. 1964. A neutral, calcium-activated proteinase from the soluble fraction of rat brain. *J. Biol. Chem. 239*:149-155.

51. Hackenthal, E., R. Hackenthal, and U. Hilgenfeldt. 1978. Purification and partial characterization of rat brain acid proteinase (isorenin). *Biochem. Biophys. Acta 522*:561-573.

52. Hemminki, K. 1973. Relative turnover of tubulin subunits in rat brain. *Biochim. Biophys. Acta 310*:285-288.

53. Hemminki, K. 1973. Turnover of actin in rat brain. *Brain Res. 57*:259-260.

54. Hirose, S., H. Yokosawa, and T. Inagami. 1978. Immunochemical identification of renin in rat brain and distinction from acid proteases. *Nature 274*:392-393.

55. Hirose, S., H. Yokosawa, T. Ingami, and R. J. Workman. 1980. Renin and prorenin in hog brain: Ubiquitous distribution and high concentration in the pituitary and pineal. *Brain Res. 191*:489-499.

56. Hirsch, H. E., and M. E. Parks. 1973. The quantitative histochemistry of acid proteinase in the nervous system. Localization in neurons. *J. Neurochem. 21*:453-458.

57. Hodgkin, A. L., and B. Katz. 1949. Effect of calcium on the axoplasm of giant nerve fibres. *J. Exp. Biol. 26*:292-294.

58. Hodgkin, A. L., and R. D. Keynes. 1956. Experiments on injections of substances into squid giant axons by means of a microsyringe. *J. Physiol. (Lond.) 131*:592-616.

59. Hoffman, P. N., and R. J. Lasek. 1975. The slow component of axonal transport: Identification of major structural polypeptides of the axon and their generality among mammalian neurons. *J. Cell Biol. 66*:351-366.

60. Hogan, E., N. Banik, R. Happel, and M. Sostek. 1982. Ca^{++}-mediated degradation of filament proteins in CNS. *Trans. Am. Soc. Neurochem. 13*:107 (abs.).

61. Huang, J. S., S. S. Huang, and J. Tang. 1979. Cathepsin D isozymes from porcine spleens: Large scale purification and polypeptide chain arrangements. *J. Biol. Chem. 254*:11405-11417.

62. Huston, R. B., and E. G. Krebs. 1968. Activation of skeletal muscle phosphorylase kinase by Ca^{2+}: II. Identification of the kinase activating factor as a proteolytic enzyme. *Biochemstry 7*:2116-2122.

63. Inagami, T., D. L. Clemens, M. R. Celio, A. Brown, L. Sandru, N. Herschkowitz, L. H. Hoffman, and A. G. Kasselberg. 1980. Immunohistochemical localization of renin in mouse brain. *Neurosci. Lett. 18*:91-98.

64. Inoue, M., A. Kishimoto, Y. Takai, and Y. Mishizuka. 1977. Studies on a cyclic nucleotide-independent protein kinase and its proenzyme in mammalian tissues: II. Proenzyme and its activation by calcium dependent protease from rat brain. *J. Biol. Chem. 252*:7610-7616.

65. Ishiura, S. 1981. Calcium-dependent proteolysis in living cells. *Life Sci.* 29:1079-1087.

66. Ishiura, S., H. Murofushi, K. Suzuki, and K. Imahori. 1978. Studies of a calcium-activated neutral protease from chicken skeletal muscle: I. Purification and characterization. *J. Biochem.* 84:225-230.

67. Ishiura, S., H. Sugita, K. Suzuki, and K. Imahori. 1979. Studies of calcium-activated neutral protease from chicken skeletal muscle: II. Substrate specificity. *J. Biochem.* 86:579-581.

68. Joys, T. M., and K. Kim. 1979. The susceptibility to tryptic hydrolysis of peptide bonds involving ε-N-methyllysine. *Biochim. Biophys. Acta* 581:360-362.

69. Kalish, F., N. Chovick, and J. F. Dice. 1979. Rapid *in vivo* degradation of glycoproteins isolated from cytosol. *J. Biol. Chem.* 254:4445-4481.

70. Katunuma, N. 1977. New intracellular proteases and their role in intracellular enzyme degradation. *Trends Biochem. Sci.* 2:122-125.

71. Kishimoto, A., N. Kajikawa, H. Tabuchi, M. Shiota, and Y. Nishizuka. 1981. Calcium-dependent neutral proteases, widespread occurrence of a species of protease active at lower concentrations of calcium. *J. Biochem.* 90:889-892.

72. Knowles, S., and F. Ballard. 1976. Selective control of the degradation of normal and aberrant proteins in Reuber H 35 hepatoma cells. *Biochem. J.* 156:609-617.

73. Kominami, E., and N. Katunuma. 1982. Immunological studies on cathepsins B and H from rat liver. *J. Biochem.* 91:67-71.

74. Komiya, Y., and M. Kuraokawa. 1979. Preferential blockade of the tubulin transport by colchicine. *Brain Res.* 190:505-516.

75. Krieger, D. T., and A. S. Liotta. 1979. Pituitary hormones in brain: Where, how, and why? *Science* 205:366-371.

76. Krystosek, A., and N. W. Seeds. 1981. Plasminogen activator secretion by granule neurons in cultures of developing cerebellum. *Proc. Natl. Acad. Sci. USA* 78:7810-7814.

77. Kubota, S.-I., K. Suzuki, and K. Imahori. 1981. A new method for the preparation of a calcium activated neutral protease highly sensitive to calcium ions. *Biochem. Biophys. Res. Commun.* 100:1189-1194.

78. Lasek, R. J., and M. M. Black. 1977. How do axons stop growing? Some clues from the metabolism of the proteins in the slow component of axonal transport. *In*: Mechanisms, Regulations, and Special Functions of Protein Synthesis of the Brain. S. Roberts, A. Lajtha, and W. H. Gispen, editors. Elsevier/North Holland, Biomedical Press, Amsterdam, pp. 161-169.

79. Lasek, R. J., and P. N. Hoffman. 1976. The neuronal cytoskeleton, axonal transport and axonal growth. *In*: Cell Motility. Vol. 3. R. Goldman, T. Pollard, and J. Rosenbaum, editors. Cold Spring Harbor Laboratory, Cold Spring Harbor, N.Y., pp. 1021-1049.

80. Lasek, R. J., N. Krishnan, and I. R. Kaiserman-Abramof. 1979. Identification of the subunit proteins of 10-nm neurofilaments isolated from axoplasm of squid and *Myxicola* giant axons. *J. Cell Biol.* 82:336-346.

81. LaVail, J. H., and M. M. LaVail. 1974. The retrograde intraaxonal transport of horseradish peroxidase in the chick visual system: A light and electron microscopic study. *J. Comp. Neurol.* 157:303-358.

82. Llinas, R., J. R. Blinks, and C. Nicholson. 1972. Calcium transient in presynaptic terminal of squid giant synapse: Detection with aequorin. *Science* 176:1127-1129.

83. Malik, M. N., L. A. Meyers, K. Iqbal, A. M. Sheikh, L. Scotto, and H. M. Wisniewski. 1981. Calcium activated proteolysis of fibrous proteins in central nervous system. *Life Sci.* 29:795-802.

84. Marcano de Cotte, D., C. E. L. De Menezes, G. W. Bennett, and J. A. Edwardson. 1980. Dopamine stimulates the degradation of gonadotropin releasing hormone by rat synaptosomes. *Nature* 283:487-489.

85. Marks, N. 1976. Biodegradation of hormonally active peptides in the central nervous

system. *In*:Subcellular Mechanisms in Reproductive Neuroendocrinology. F. Naftolin, K. J. Ryan and A. J. Davies, editors. Elsevier, Amsterdam, pp. 129-147.

86. Marks, N., and A. Lajtha. 1965. Separation of acid and neutral proteinases of brain. *Biochem. J.* 97:74-83.

87. Marks, N., and A. Lajtha. 1971. Protein and polypeptide breakdown. *In*: Handbook of Neurochemistry. Vol. 5. A. Lajtha, editor. Plenum Press, New York, pp. 49-139.

88. McLendon, G., and E. Radany. 1978. Is protein turnover thermodynamically controlled? *J. Biol. Chem.* 253:6335-6337.

89. Mego, J., and R. Farb. 1978. An energy requirement for the degradation of intravenously injected ^{125}I-labelled albumin in mouse liver and kidney slices. *Biochem. J.* 172: 233-238.

90. Mellgren, R. L. 1980. Canine cardiac calcium-dependent proteases: Resolution of two forms with different requirements for calcium. *FEBS Lett.* 109:129-133.

91. Mellgren, R. L., J. H. Aylward, S. D. Killilea, and E. Y. C. Lee. 1979. The activation and dissociation of a native high molecular weight form of rabbit skeletal muscle phosphorylase phosphatase by endogenous Ca^{2+}-dependent proteases. *J. Biol. Chem.* 254:648-652.

92. Morris, J. R., and R. J. Lasek. 1980. Metabolic activities associated with the stable cytoskeleton of axoplasm. *J. Cell Biol.* 87:1304.

93. Mort, J. S., A. R. Poole, and R. S. Decker. 1981. Immunofluorescent localization of cathepsins B and D in human fibroblasts. *J. Histochem. Cytochem.* 29:649-657.

94. Mortimore, G. E., K. H. Woodside, and J. E. Henry. 1972. Compartmentation of free valine and its relation to protein turnover in perfused rat liver. *J. Biol. Chem.* 247:2776-2784.

95. Murakami, T., M. Hatanaka, and T. Murachi. 1981. Calpain I and calpastatin in human erythrocytes. *J. Biochem.* 90:1810-1816.

96. Neely, A., P. Nelson, and G. Mortimore. 1974. Osmotic alterations of the lysosomal system during rat liver perfusion: Reversible suppression by insulin and amino acids. *Biochim. Biophys. Acta* 338:458-472.

97. Neurath, H., and K. A. Walsh. 1976. Role of proteolytic enzymes in biological regulation (a review). *Proc. Natl. Acad. Sci. USA* 73:3825-3832.

98. Nixon, R. A. 1980. Protein degradation in the mouse visual system: I. Degradation of axonally transported and retinal proteins. *Brain Res.* 200:69-83.

99. Nixon, R. A. 1982. Increased axonal proteolysis in myelin deficient mutant mice. *Science* 215:999-1001.

100. Nixon, R. A., B. A. Brown, and C. A. Marotta. 1983. Limited proteolytic modification of a neurofilament protein involves a proteinase activated by endogenous levels of calcium. *Brain Res.*, in press.

101. Nixon, R. A., B. A. Brown, and C. A. Marotta. 1982. Post-translational modification of a neurofilament protein during axoplasmic transport: Implications for regional specialization of CNS axons. *J. Cell Biol.* 94:150-158.

102. Nixon, R. A., and M. Froimowitz. 1982. Multiple neutral proteinases in retinal ganglion cell neurons. *Trans. Am. Soc. Neurochem.* 13:238 (abs.).

103. Orlowski, M., C. Michaud, and S. Wilk. 1980. Generation of methionine and leucine-enkephalin from precursor molecules by cation-sensitive neutral endopeptidase of bovine pituitary. *Biochem. Biophys. Res. Commun.* 94:1145-1153.

104. Orlowski, M., E. Wilk, S. Pearce, and S. Wilk. 1979. Purification and properties of a prolyl endopeptidase from rabbit brain. *J. Neurochem.* 33:461-469.

105. Orrego, F. 1971. Protein degradation in squid giant axons. *J. Neurochem.* 18:2249-2254.

106. Osman, M. Y., R. R. Smeby, and S. Sen. 1979. Separation of dog brain renin-like activity from acid protease activity. *Hypertension* 1:53-59.

107. Palay, S. L. 1958. The morphology of synapses in the central nervous system. *Exp. Cell Res.* 5:275-293.

108. Pant, H. C., and H. Gainer. 1980. Properties of a calcium-activated protease in squid axoplasm which selectively degrades neurofilament proteins. *J. Neurobiol. 11*:1-12.

109. Pant, H. C., H. B. Pollard, G. D. Pappas, and H. Gainer. 1979. Phosphorylation of specific, distinct proteins in synaptosomes and axons from squid nervous system. *Proc. Natl. Acad. Sci. USA 76*:6071-6075.

110. Pant, H. C., S. Terakawa, and H. Gainer. 1979. A calcium activated protease in squid axoplasm. *J. Neurochem. 32*:99-102.

111. Paskin, N., and R. J. Mayer. 1978. A method for the anlaysis of protein turnover characteristics: Indirect estimation of rates of protein degradation. *Biochem. J. 174*:153-161.

112. Pennington, R. J. T. 1977. Proteinases of muscle. *In*: Proteinases in Mammalian Cells and Tissues. A. J. Barrett, editor. Elsevier/North-Holland Biomedical Press, Amsterdam, pp. 515-543.

113. Peters, A., S. L. Palay, and H. deF. Webster. 1970. *In*: The Fine Structure of the Nervous System: The Cells and Their Processes. Harper and Row, New York, pp. 1-198.

114. Puca, G. A., E. Nola, V. Sica, and F. Bresciani. 1977. Estrogen binding proteins of calf uterus: Molecular and functional characterization of the receptor transforming factor: A Ca^{++}-activated protease. *J. Biol. Chem. 252*:1358-1366.

115. Puizdar, V., and V. Turk. 1981. Cathepsinogen D: Characterization and activation to cathepsin D and inhibitory peptides. *FEBS Lett. 132*:299-304.

116. Reddy, M. K., J. D. Etlinger, M. Rainowitz, D. A. Fischman, and R. Zak. 1975. Removal of Z-lines and alpha-actinin from isolated myofibrils by a calcium-activated neutral protease. *J. Biol. Chem. 250*:4278-4284.

117. Requena, J., and L. J. Mullins. 1979. Calcium movement in nerve fibres. *Quart. Rev. Biophys. 12*:371-460.

118. Riekkinen, P. J., and U. K. Rinne. 1968. A new neutral proteinase from the rat brain. *Brain Res. 9*:126-135.

119. Rupnow, J. H., W. L. Taylor, and J. E. Dixon. 1979. Purification and characterization of a thyrotropin-releasing hormone deamidase from rat brain. *J. Biol. Chem. 18*:1206-1211.

120. Sandberg, M., A. Hamberger, I. Jacobson, and J-O. Karlsson. 1980. Role of calcium ions in the formation and release of low-molecular-weight substances from optic nerve terminals. *Neurochem. Res. 5*:1185-1198.

121. Sandoval, I. V., and K. Weber. 1978. Calcium-induced inactivation of microtubule formation in brain extracts: Presence of a calcium-dependent protease acting in polymerization-stimulating microtubule-associated proteins. 1978. *Eur. J. Biochem. 92*:463-470.

122. Sapolsky, A. I., and J. F. Woessner, Jr. 1972. Multiple forms of cathepsin D from bovine uterus. *J. Biol. Chem. 247*:2060-2076.

123. Schally, A. V., W. Huang, R. C. C. Chang, A. Arimura, T. Redding, R. Millar, M. W. Hunkapiller, and L. E. Hood. 1980. Isolation and structure of pro-somatostatin: A putative somatostatin precursor from pig hypothalamus. *Proc. Natl. Acad. Sci. USA 77*:4489-4493.

124. Schimke, R. T., and M. O. Bradley. 1975. Properties of protein turnover in animal cells and a possible role for turnover in "quality" control of proteins. *In*: Proteases and Biological Control. Vol. 2. E. Reich, D. B. Rifkin, and E. Shaw, editors. Cold Spring Harbor Laboratory, Cold Spring Harbor, N.Y., pp. 515-531.

125. Schlaepfer, W. W. 1979. Nature of mammalian neurofilaments and their breakdown by calcium. *In*: Progress in Neuropathology. Vol. 4. H. M. Zimmerman, editor. Raven Press, New York, pp. 101-123.

126. Schlaepfer, W. W., and L. A. Freeman. 1980. Calcium-dependent degradation of mammalian neurofilaments by soluble tissue factor(s) from rat spinal cord. *Neuroscience 5*: 2305-2314.

127. Schlaepfer, W. W., and M. B. Hasler. 1979. Characterization of the calcium-induced disruption of neurofilaments in rat peripheral nerve. *Brain Res. 168*:299-309.

128. Schlaepfer, W. W., and S. Micko. 1979. Calcium-dependent alterations of neurofilament proteins of rat peripheral nerve. *J. Neurochem. 32*:211-219.

129. Schlaepfer, W. W., and U.-J. P. Zimmerman. 1981. Calcium-activated proteolysis of neurofilament protein. *Inter. Soc. Neurochem. (Abstr.) 8*:20.

130. Schlaepfer, W. W., and U.-J. P. Zimmerman. 1981. Calcium-mediated breakdown of glial filaments and neurofilaments in rat optic nerve and spinal cord. *Neurochem. Res. 6*: 243-255.

131. Schlaepfer, W. W., U.-J. P. Zimmerman, and S. Micko. 1981. Neurofilament proteolysis in rat peripheral nerve: Homologies with calcium-activated proteolysis of other tissues. *Cell Calcium 2*:235-250.

132. Sinha, A. K., and S. P. R. Rose. 1972. Compartmentation of lysosomes in neurones and neuropil and a new neuronal marker. *Brain Res. 39*:181-196.

133. Skene, J. H. P., and M. Willard. 1981. Characteristics of growth-associated polypeptides in regenerating toad retinal ganglion cell axons. *J. Neurosci. 1*:419-426.

134. Smith, R., and V. Turk. 1974. Cathepsin D: Rapid isolation by affinity chromatography on methemoglobin-agarose resin. *Eur. J. Biochem. 48*:245-254.

135. Sommercorn, J. M., and R. W. Swick. 1981. Protein degradation in primary monolayer cultures of adult rat hepatocytes. *J. Biol. Chem. 256*:4816-4821.

136. Soreq, H., and R. Miskin. 1981. Plasminogen activator in the rodent brain. *Brain Res. 216*:361-374.

137. Stromska, D. P., and S. Ochs. 1981. Patterns of slow transport in sensory nerves. *J. Neurobiol. 12*:441-453.

138. Sugita, H., S. Ishiura, K. Suzuki, and K. Imahori. 1980. Ca^{++}-activated neutral protease and its inhibitors: In vitro effect on intact myofibrils. *Muscle & Nerve 3*:335-339.

139. Suhar, A., and N. Marks, 1979. Purification and properties of brain cathepsin B. *Eur. J. Biochem. 101*:23-30.

140. Suzuki, K., S. Ishiura, S. Tsuji, T. Katamoto, H. Sugita, and K. Imahori. 1979. Calcium-activated neutral protease from human skeletal muscle. *FEBS Lett. 104*:355-358.

141. Suzuki, K., S. Tsuji, S. Ishiura, Y. Kimura, S. Kubota, and K. Imahori. 1981. Autolysis of calcium-activated neutral protease of chicken skeletal muscle. *J. Biochem. 90*: 1787-1793.

142. Suzuki, K., S. Tsuji, S. Kubota, Y. Kimura, and K. Imahori. 1981. Limited autolysis of Ca^{2+}-activated neutral protease (CANP) changes its sensitivity to Ca^{2+} ions. *J. Biochem. 90*:275-278.

143. Takahashi, S., K. Murakami, and Y. Miyake. 1981. Purification and characterization of porcine kidney cathepsin B_1. *J. Biochem. 90*:1677-1684.

144. Takahashi-Nakamura, M., S. Tsuji, K. Suzuki, and K. Imahori. 1981. Purification and characterization of an inhibitor of calcium-activated neutral protease from rabbit skeletal muscle. *J. Biochem. 90*:1583-1589.

145. Tillotson, D., and A. L. F. Gorman. 1980. Non-uniform Ca^{2+} buffer distribution in a nerve cell body. *Nature 286*:816-817.

146. Truglia, J. A., and A. Stracher. 1981. Purification and characterization of a calcium dependent sulfhydryl protease from human platelets. *Biochem. Biophys. Res. Commun. 100*:814-822.

147. Tsuji, S., and K. Imahori. 1981. Studies on the Ca^{2+}-activated neutral proteinase of rabbit skeletal muscle: I. The characterization of the 80 K and the 30 K subunits. *J. Biochem. 90*:233-240.

148. Ueno, N., H. Miyazaki, S. Hirose, and K. Murakami. 1981. A 56,000-dalton renin-binding protein in hog kidney is an endogenous renin inhibitor. *J. Biol. Chem. 256*:12023-12027.

149. Vedeckis, W. V., M. R. Freeman, W. T. Schrader, and B. W. O'Malley. 1980. Progesterone-binding components of chick oviduct: Partial purification and characterization of

a calcium-activated protease which hydrolyzes the progesterone receptor. *Biochemistry 19*: 335-343.

150. 'Ward, W. F., J. R. Cox, and G. E. Mortimore. 1977. Lysosomal sequestration of intracellular protein as a regulatory step in hepatic proteolysis. *J. Biol. Chem. 252*:6955-6961.

151. Ward, W., and G. Mortimore. 1978. Compartmentation of intracellular amino acids in rat liver, evidence for an intralysosomal pool derived from protein degradation. *J. Biol. Chem. 253*:3581-3587.

152. Wasman, L., and E. G. Krebs. 1978. Identification of two protease inhibitors from bovine cardiac muscle. *J. Biol. Chem. 253*:5888-5891.

153. Weiss, P. A., and H. B. Hiscoe. 1948. Experiments on the mechanism of nerve growth. *J. Exp. Zool. 107*:315-395.

154. Whitaker, J. N., and J. M. Seyer. 1979. Isolation and characterization of bovine brain cathepsin D. *J. Neurochem. 32*:325-333.

155. Whitaker, J. N., and J. M. Seyer. 1981. The influence of pH on the degradation of bovine myelin basic protein by bovine brain cathepsin D. *Biochim. Biophys. Acta 661*:334-341.

156. Whitaker, J. N., L. C. Terry, and W. O. Whetsell, Jr. 1981. Immunocytochemical localization of cathepsin D in rat neural tissue. *Brain Res. 216*:109-124.

157. Wilk, S., and M. Orlowski. 1979. Degradation of bradykinin by isolated neutral endopeptidases of brain and pituitary. *Biochem. Biophys. Res. Commun. 90*:1-6.

158. Wilk, S., S. Pearce, and M. Orlowski. 1979. Identification and partial purification of a cation-sensitive neutral endopeptidase from bovine pituitaries. *Life Sci. 24*:457-464.

159. Willard, M., M. Wiseman, J. Levine, and P. Skene. 1979. Axonal transport of actin in rabbit retinal ganglion cells. *J. Cell Biol. 81*:581-591.

160. Wood, J. G., and N. King. 1971. Turnover of basic protein of rat brain. *Nature 229*:56-57.

161. Yoshimoto, T., W. H. Simmons, T. Kita, and D. Tsuru. 1981. Post-proline cleaving enzyme from lamb brain. *J. Biochem. 90*:325-334.

162. Zak, R., A. F. Martin, and R. Blough. 1979. Assessment of protein turnover by use of radioisotopic tracers. *Physiol. Rev. 59*:407-447.

163. Zimmerman, U-J. P., and W. W. Schlaepfer, 1982. Characterization of calcium-activated protease from rat brain and skeletal muscle using affinity chromatography with ^{125}I-neurofilament proteins as ligand. *Fed. Proc. 41*:1387 (abs.).

Chapter 6

Experimental Models
of Neurofilamentous Pathology

Kenneth S. Kosik and Dennis J. Selkoe

The development of experimental systems in which the cellular economy of neurofilaments is altered has importance to two broad areas of investigation. First, such models provide tissue in which specific aspects of the cell biology of normal neurofilaments (for example, synthesis, transport, degradation) can be manipulated and thus be more readily studied than in normal neurons. At the same time, they may provide abundant quantities of starting material for certain structural and physicochemical analyses of *in situ* or purified neurofilaments. Second, the use of animal and tissue culture systems allows one to derive dynamic information about the sequence of cellular and molecular events that leads to pathological accumulation or reorganization of neuronal intermediate filaments. Insights about the mechanisms of filamentous degeneration of neurons and axons can be obtained that may be relevant to neurofibrillary disease in humans, which can necessarily be studied only in a static fashion.

Interest in the use of neurofilamentous models for the latter purpose has increased dramatically during the past two decades with the realization that filamentous degeneration of neurons and their axonal terminals represents the most common neuropathological change in the brains of aged humans with progressive intellectual failure. The marked and continuing rise in the portion of the population surviving beyond the sixth decade and the fact that fibrillary lesions are the most characteristic age-related morphological alterations even in the brains of the normal elderly have futher spurred interest in the role of the neuronal cytoskeleton in aging and disease. In addition, numerous genetic, toxic, or idiopathic neurological disorders have been found in recent years to be characterized by striking accumulations of neurofilaments in axons and/or neuronal cell bodies of the peripheral and central nervous systems. Although individually less frequent than senile dementia of the Alzheimer type, these varied human diseases emphasize the urgency of understanding the mechanisms of pathological organization of neuronal filaments. Thus, on the basis of both fundamental and clinical considerations, the development and analysis of experimental models of filamentous pathology represent an important area of neurobiological investigation.

General Principles

Even a cursory review of the rapidly expanding literature on neurofilamentous pathology leaves the reader impressed by the great diversity of experimental manipulations that may lead to rather similar patterns of filamentous degeneration in axons and/or perikarya. As will be seen in the following pages, many chemically unrelated toxins, certain disorders of intermediary metabolism, some autosomally inherited diseases, and occasional traumatic lesions can all be associated with morphologically similar reorganizations of neuronal intermediate filaments. In this regard, two points are worthy of emphasis. First, patterns of filamentous change that appear the same at the level of the light microscope and initially even by electron microscopy may, on detailed reexamination, demonstrate differences in the sequence of development of the lesions over time or in the relationship of the reorganized filaments to other altered organelles in the cytoplasm. The experimental conditions that induce prominent filamentous lesions affect other neuronal organelles as well, and the latter changes may provide important clues to fundamental differences among filamentous pathological processes. Second, even two conditions that produce morphologically indistinguishable filamentous lesions may have distinct pathogeneses, since filament reorganization may be a nonspecific, end-stage change that can be caused by a variety of distinct, highly specific defects in neuronal metabolism.

A few years ago, available evidence suggested that certain filament-reorganizing toxins selectively induced changes in neurons of peripheral nervous system while others might principally affect central nervous system cells. With more extensive neuropathological study of such agents, it has become increasingly apparent that this distinction cannot be rigorously applied to most toxins. The route of administration rather than the chemical nature of the toxin itself will usually dictate whether peripheral or central neurons are predominantly affected. This is true, for example, of the mitotic spindle inhibitor, vincristine, which can be injected intrathecally to induce neurofilament bundles in spinal neuronal perikarya and intraneurally to induce axonal swellings in peripheral nerves.

It is often useful to take advantage of such topographical selectivity of the same toxin administered by different routes in designing experiments that address various questions about neurofilament economy. In the authors' work, intrathecal injection of aluminum ions was used to induce large numbers of neurofilament bundles in many spinal neurons in order to harvest filament-containing cells quantitatively and to examine the polypeptide composition of the aluminum-induced filaments (158). When we turned our attention to abnormalities of axonal transport in the aluminum model, we realized that intrathecal administration induces filamentous lesions in only a minor percentage of the large motor neurons in the periphery of the anterior horn (158) and thus is not well suited to the study of transport in the sciatic nerve axons. Instead, we employed selective lumbar intraspinal injection of microaliquots of concentrated aluminum chloride, a technique that results in a very high proportion of large motor neurons being affected (95). Moreover, this model has an additional advantage for transport studies: rabbits with extensive pathology can survive for weeks or months in contrast to the early death that usually ensues within days of the appearance of neurological signs following intrathecal aluminum administration. The direct application of IDPN (61,120) and n-hexane (61) to exposed peripheral nerves, rather than systemic administration, in order to detect early local effects on neurofilaments prior to widespread distribution and metabolism of the toxins provides another example of the usefulness of varying the route of administration in mechanistic studies.

The relative involvement of neuronal cell bodies, proximal axons, or distal axons in various models of neurofilamentous pathology can also be exploited in designing experiments that address questions about the subcellular distribution and local processing of neurofilaments. Several chemical agents, including aliphatic hexacarbons, acrylamide, and carbon disulfide, induce focal accumulations of normal-appearing neurofilaments and other transported organelles in the distal portions of long and large-diameter axons of both central and peripheral neurons (171). These swellings occur at progressively more proximal loci during the course of the intoxication; as a result, this process has been referred to as dying-back axonal degeneration, or more recently, as central-peripheral distal axonopathy (Figure 6-1). Although it was originally believed that axonal terminals were the site of the first lesions, it has been shown that the initial swellings arise at the distal end of the main nerve fiber, proximal to the thinner branches supplying the terminals, which develop neurofilament swellings subsequently (175).

In contrast to this distal pattern, the lathyrogenic neurotoxin, β, β'-iminodipropionitrile (IDPN), induces neurofilament-rich swellings in proximal axons. This model was the first experimental neurofibrillary

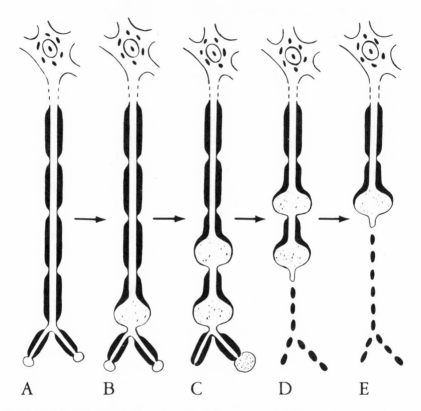

Figure 6-1. Distal (dying-back) axonopathy produced by neurotoxic hexacarbon compounds such as MnBK and 2,5-HD. (A) Normal axon. During chronic intoxication, transported materials (stippling) accumulate, causing giant axonal swellings and myelin retraction to develop on the proximal sides of nodes of Ranvier in the distal portion of the fiber (B). Paranodal swellings appear at more proximal loci with time as transport failure develops at these sites (C), while the distal portion below an axonal swelling undergoes Wallerian-like degeneration (D and E). (From reference 171. Used with permission.)

degeneration for which a specific mechanism of filament accumulation was shown. Griffin and associates (63) found that IDPN-induced proximal swellings resulted from a selective blockade of the transport of proteins in the slow component of axoplasmic flow, particularly the neurofilament polypeptides. Recently, these investigators demonstrated that aluminum, administered intrathecally, also causes filaments to accumulate initially in the proximal axon rather than in the perinuclear cytoplasm of the cell body, as previously assumed (192); the role of altered filament transport in this process is not yet established. Later in the course of aluminum neurofibrillary degeneration, the histopathology is primarily perikaryal in location. This temperospatial pattern seen in the aluminum model may be relevant to the pathogenesis of the neurofilamentous masses seen in proximal axons and cell bodies of motor neurons in sporadic human motor neuron disease (amyotrophic lateral sclerosis).

In addition to the presence of filamentous lesions in neuronal perikarya and/or axons, certain models also show morphological reorganization of intermediate filaments in nonneuronal cells. This is obviously true for mitotic spindle inhibitors such as cochicine and colcemid, whose filament-aggregating properties were first defined in nonneural cells. An example of a genetically determined neurological disorder displaying a generalized abnormality of intermediate filaments has been found in the human disease giant axonal neuropathy (127). This rare, apparently recessive disease results in an ascending peripheral neuropathy with onset in childhood and progression to death over a period of some years. Focal filamentous axonal swellings similar to those in hexacarbon neuropathy are found in distal portions of peripheral nerves. Pena (127) has recently demonstrated that cultured skin fibroblasts from a patient with giant axonal neuropathy display abnormal juxtanuclear aggregates of their intermediate filaments composed of vimentin (127). A spontaneously occurring animal model of giant axonal neuropathy has been discovered in certain inbred dogs (42); distal axonal filament bundles were present in both peripheral and central axons (43,68). Although some Schwann cells also contained increased numbers of cytoplasmic filaments (43), reorganization of intermediate filaments in other nonneural cells has not yet been reported in the canine model. Further examination of this model as well as exposure of cultured fibroblasts and other intermediate filament-containing, nonneural cells to the various neurofilament-inducing toxins should prove to be very useful in studying the pathogenesis of generalized disorders of intermediate filament organization.

Investigators contemplating the study of filament-inducing neurotoxins in animals should also be aware of the marked interspecies variability in the effects of some agents. Perhaps the most striking example is the high degree of susceptibility of rabbits, cats, and guinea pigs to the filament-reorganizing action of aluminum, in contrast to the relative resistance of rats and rhesus monkeys to this metal (34). For example, King and associates (90) found that two strains of rats developed neither neurofibrillary lesions nor a clinical encephalopathy at brain aluminum concentrations of up to 40 μg per gram, a level 5 to 10 times that inducing widespread, lethal brain degeneration in the rabbit or cat. Most of the other experimental neurotoxins do not show as marked interspecies differences as does aluminum.

In the following pages, we will review the important features of a wide variety of experimental manipulations known to alter the organization and/or quantity of neuronal intermediate filaments. Where possible, we will discuss the current understanding of the mechanisms of filament reorganization and will describe the variety of ways in which a particular model or toxin has been employed in animals or in tissue cultures. We have not included all experimental conditions that may be associated with neurofilamentous alteration. Conditions in which a relatively minor degree of accumulation, degeneration, or rearrangement of neurofilaments occurs in association with numerous other cytopathological changes have not been considered here. Rather, we have concentrated on models in which neurofilamentous alteration appears to represent the predominant structural event in the pathogenesis of the neuronal or axonal degeneration. We have also addressed only those studies that provide information of direct relevance to morphologists, biochemists, and cell biologists interested in neuronal intermediate filaments in normal and pathological states. A considerable number of papers dealing with behavioral, electrophysiological, and biochemical changes in animals displaying neurofibrillary degeneration have not been reviewed here because their relevance to the altered neurofilaments remains unclear. For this information, the reader is referred to available monographs on the general subjects of neurotoxicology or animal models in neurological disease.

Aluminum

Aluminum as an agent inducing neurofibrillary accumulation has attracted wide attention. This interest arises not only because of the element's remarkable ability to induce massive accumulations of neurofilaments within perikarya and axons but also because of the report of a selective increase in aluminum in neurofibrillary tangle-containing neurons of some patients with Alzheimer's disease (129). Although the relation between aluminum and Alzheimer's disease requires further elucidation, aluminum as a neurotoxin has great heuristic value in its own right. The neurotoxicity of aluminum was realized by a number of early investigators, some even before the turn of the century (39,151,155,166). Aluminum, in the form of discs implanted under the dura, was applied to brain tissue by Kopeloff and others (93), who documented its epileptogenic activity. In 1965, Klatzo, Wisniewski, and Streicher (92) first described the presence of neurofibrillary lesions in aluminum-treated animals. This fortuitous finding resulted from the use of Holt's adjuvant, which contains alum phosphate, for some otherwise unrelated immunochemical experiments.

Subsequently, it was realized that aluminum ions administered in any stable chemical form could induce neurofilamentous changes.

The most common methods of administration are either intracisternal or intraparenchymal. Both methods are effective; the latter produces a more focal topography of the lesions. When brain or spinal cord is studied by light microscopy following full development of the lesions, it is apparent that vascular and glial elements are spared and that the lesions are specifically confined to certain neurons. At this stage, the lesions consist of perikaryal neurofibrillary tangles demonstrable with silver stains or with the dye Congo red under polarized light. In the affected neurons, the normal neurofibrillary meshwork is replaced by large, dense, fibrous tangles that appear as discrete bundles with sharp borders but without limiting membranes. In some neurons, the tangles become so large that the neuron is completely filled and has a homogeneous hyalinized appearance (Figure 6-2). Normal neuronal structures such as Nissl substance and the nucleus are occasionally seen to be displaced into a small area by the massive fibrillary bundles (Figure 6-3). Nissl substance is thought to be displaced rather than replaced because the RNA content of cells with filamentous tangles is not reduced (46). Outside the areas with neurofibrillary tangles, the neuronal architecture is not greatly altered. An exception may be dendritic morphology, concerning which Petit and others (132) have reported a dendritic dying back process in layer V pyramidal cells of sensorimotor cerebral cortex in aluminum-treated rabbits. These authors noted a sharp and progressive decrease in the number of dendritic branches with increasing distance from the cell body.

Two lines of evidence suggest that the aluminum-induced accumulations consist of neurofilaments. By electron microscopy, the tangles are composed of 10-nm filamentous structures (189); this finding classifies them as intermediate filaments. Biochemically, isolated rabbit spinal neurons containing massive accumulations of filaments have been shown to contain several-fold increases in the neurofilament triplet proteins (158). Furthermore, an antiserum raised against gel-purified 160,000-mol wt protein of bovine neurofilaments produced specific and intense fluorescent staining of aluminum-induced neurofilament bundles (158). Antineurofilament sera prepared against extracts of chicken brain and human sciatic nerve neurofilament preparations also stained these bundles (32).

Although aluminum encephalomyelopathy has been proposed as a potential model for Alzheimer neurofibrillary degeneration, it must be emphasized that the ultrastructure of Alzheimer paired helical filaments (88,205) is quite distinct from the aluminum-induced

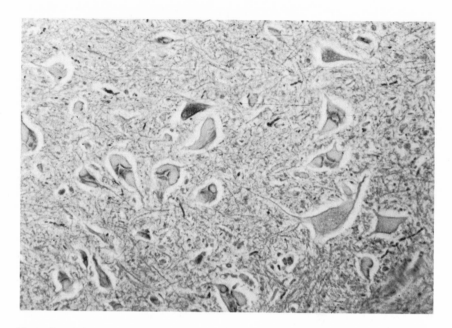

Figure 6-2. Bodian stain of the lateral motor nucleus from the lumbar spinal cord of a rabbit treated with intraspinal microinjections of aluminum chloride, as described in the text. Extensive involvement of the large anterior horn cells with neurofibrillary bundles is demonstrated.

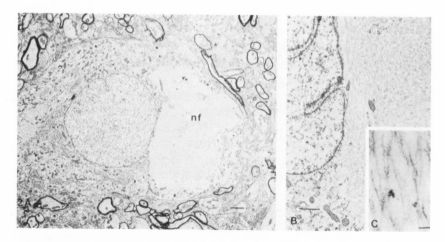

Figure 6-3. Electron microscopy of *in situ* filament-containing neurons. (A) Large perinuclear swirl of neurofilaments (nf) displaces other organelles in an anterior horn cell. (Bar represents 2 μm.) (B and C) Higher magnifications. (Bar in B represents 1 μm; bar in C represents 100 nm.) (From reference 158. Used with permission.)

straight, unpaired neurofilaments. Furthermore, the filaments comprising the paired helical filaments have not been identified as neurofilaments. In addition to their morphological differences, the two types of fibrous structures differ markedly in their solubility characteristics (157).

Much has been written about the topography of aluminum-induced changes in brain and spinal cord. The density of aluminum filamentous lesions is proportional to the concentration of brain aluminum (30,212). Utilizing bilateral hippocampal injections in cats, Crapper and Tomko (30) found that regions with aluminum concentrations less than 4 μg per gram dry weight were free of neurofibrillary tangles. However, with relatively similar aluminum concentrations, there is variability in resistance to neurofibrillary change among different neuronal populations. For example, the caudate nucleus is spared even after intraventricular injection (211). The hypoglossal and vagal nuclei, which are nearly adjacent to each other, show a marked differential susceptibility, with the former being considerably more vulnerable to the effects of aluminum (203). Even within a defined group of neurons in intoxicated animals, one neuron may show neurofibrillary tangles while an adjoining one may appear normal. The distribution of the neurofibrillary accumulations and the concentration of aluminum throughout the central nervous system depends in part upon the route of administration. Intracisternal injections cause changes predominantly in spinal cord, to a lesser extent in brainstem, and only minimally in cerebral cortex. Intraparenchymal injections result in a more focal pattern of neurofibrillary lesions. Direct cerebral neocortical injections affect primarily pyramid-shaped cells of layers III and V (28,29). Intravitreal injections of 1% aluminum chloride produce neurofibrillary tangles in ganglion, bipolar, and photoreceptor cells (208). More detailed descriptions of the gross topography of aluminum-induced tangles have been presented in several papers (92,132,203,206,212).

Neurofibrillary lesions can also be induced by systemic administration of soluble aluminum salts to susceptible animals. Aluminum lactate or aluminum tartrate administered by repeated subcutaneous injections (34) leads to filamentous lesions in brain. Feeding aluminum chloride to pregnant rabbits results in neurofibrillary changes in the brains of their offspring (103). These studies show that aluminum ions can cross the blood-brain barrier. Furthermore, passage of aluminum ions across the blood-brain barrier is not the result of apparent damage to the barrier, since in the original work of Klatzo and associates (92) it was demonstrated by intravenous injection of fluorescein before sacrifice that this barrier remained intact.

The pathogenetic sequence of aluminum-induced tangles has been carefully documented by Troncoso and others (192). Following intracisternal injections of aluminum chloride, rabbits were sacrificed at various intervals. At the earliest time points, 1 and 2 days after injection, accumulations of neurofilaments were limited to the most proximal portion of the axon within the gray matter of the spinal cord. Neuronal perikarya only rarely contained such accumulations early on, and, when present, these aggregates were located at the base of larger dendrites. These recent findings raise questions about the previous distinction of aluminum neurotoxicity from IDPN neurotoxicity (see below), that is, that the former represented a perikaryal accumulation, while the latter caused proximal axonal swellings. By 4 days after aluminum injection, neurofibrillary material appeared in the perikarya, and at later stages (4 and 8 weeks postinjection) the predominant sites of involvement were perikaryal and dendritic. Although axonal filamentous swellings were less frequently seen in the later stages, those axons still containing such swellings reached enormous dimensions. These findings also lend further support to the original observation of Wisniewski and others (206) that aluminum-induced neurofibrillary lesions can be reversible. In developing rabbits, these workers observed that the aluminum-induced changes are not associated with the status epilepticus that develops in mature rabbits and results in their demise. The developing rabbits could, therefore, be maintained for up to 100 days postinjection. Such chronically surviving animals showed many fewer neurons with filamentous tangles than rabbits sacrificed shortly after injection (206).

A frequent observation among investigators is the variation among species in susceptibility to aluminum. Rabbits and cats are used in most experiments because of their marked susceptibility; mice and rats are quite resistant. The question arises, therefore, as to how vulnerable human neurons are to the effects of aluminum, particularly since aluminum has been postulated to play a role in certain human neurological disorders. De Boni and others (36) examined this question in human fetal neurons cultured from dorsal root ganglia and cerebral cortex. The dorsal root ganglionic neurons were unaffected by aluminum for periods exceeding 3 weeks even when exposed to 10 times the concentration of aluminum lethal to rabbits and cats. However, neurons cultured from cortical slabs, when exposed to aluminum at relatively low concentrations, responded with an accumulation of 10-nm filaments morphologically identical to normal neurofilaments and the filaments seen in the rabbit and cat aluminum models. These authors point out that the aluminum concentrations sufficient to induce these changes are no more than those that have been reported in the cortical neurons of Alzheimer patients. As noted earlier, however, the ultrastructure and biochemical properties of the filaments in Alzheimer's disease are clearly distinguishable from those of normal neurofilaments. Most recently, Sato and others (148) treated newborn mouse dorsal root ganglia in culture with aluminum lactate for 3 to 9 days and found an extensive and diffuse proliferation of neurofilaments in the cytoplasm. The morphology differed from that induced by mitotic spindle inhibitors, which resulted in neurofibrillary rings confined to the perinuclear region. The axons of these aluminum-exposed cells were unaffected.

A number of studies provide information about the nature and subcellular locus of the toxicity of aluminum in neurons. Using the Solochrome Azurine method, Klatzo and others (92) were able to detect characteristic blue staining for aluminum in the vicinity of the birefringent neurofibrillary tangles. The increased aluminum content of affected neurons has been confirmed by electron-probe analyses (189). Using the fluorescent stain morin, which chelates aluminum (and other metals not found in brain), De Boni and others (35) were able to localize the site of aluminum binding to neuronal nuclei. Many other cell types, including ependymal cells, oligodendrocytes, pericytes, capillary endothelial cells, neutrophils, and cells of the choroid plexus, exhibited fluorescent staining for aluminum. However, only the neurons showed the characteristic cytopathological changes. The pattern of fluorescence observed within nuclei by 6 hours after intrathecal injection suggested that aluminum ions were bound to chromatin. This suggestion of an aluminum-nucleic acid complex led to experiments in which aluminum was reacted with calf thymus DNA under various conditions. Karlik and others (87) demonstrated that aluminum (administered as the chloride salt) appeared to bind relatively strongly to native DNA since it displaced ethidium bromide, which binds preferentially to intercalation sites on the DNA double helix. When the aluminum was subsequently chelated, complete return of the ethidium bromide fluorescence occurred. Also, aluminum binding did not detectably alter such optical characteristics of the DNA molecule as its A_{260}, ultraviolet spectra, or circular dichroism. These findings imply that aluminum bound reversibly to double helical DNA without disturbing hydrogen binding. These investigators also studied the effects of aluminum on unfolding and renaturation of DNA under various conditions (87). At high pH the addition of aluminum tended to stabilize DNA, whereas at low pH destabilization occurred. At high pH, the aluminum presumably binds to DNA phosphate moieties, while at low pH the aluminum is postulated to form cross-links between the strands that permit regeneration of a double helix. This complex pH-dependent behavior is thought to result from the existence of multiple aluminum species in aqueous solution, the quantity and nature of which depend on pH and aluminum concentration (6).

An additional indication that DNA damage may occur from aluminum exposure is the effect on sister chromatid exchanges in human lymphocytes treated with aluminum (36). Sister chromatid exchanges, calculated as the number of exchange points divided by the number of chromosomes on that spread, were decreased with aluminum concentrations from 3.3 to 13.2 μg aluminum per milliliter of media. However, with higher concentrations, the number of sister chromatid exchanges returned to control values. The authors point out that this bimodal response of sister chromatid exchanges to aluminum concentration highlights the specificity of aluminum's effect on chromatin. Futher support for this concept stems from the observation of an altered puffing pattern in polytene chromosomes given an aluminum challenge (147).

The complex molecular events between aluminum-DNA binding and the production of neurofibrillary accumulations remain to be elucidated. It is known that there is an increase of the total organic mass of the neuron (46) as well as an increase of protein synthesis (31,47). An increase in [^{3}H] leucine incorporation was observed in purified neuronal fractions isolated on sucrose gradients after *in vivo* labeling (47). Paradoxically, no increase in RNA content has been detected (46). Similar results were obtained in cultured neuroblastoma cells treated with aluminum salt (112) in which 10-nm filaments accumulated. While incorporation of radioactive leucine was significantly increased, ribosomal RNA content per cell was actually decreased. Preliminary data (36) suggest that DNA synthesis as measured by the incorporation of tritiated thymidine in astrocytes *in vitro* is increased after exposure to aluminum levels of 3.3 to 6.6 μg aluminum per milliliter of medium. The authors hypothesize that this result may imply increased DNA repair. A recent study (201) reported that in rats fed a suboptimal zinc diet the activity of brain dopamine-beta-hydroxylase and phenylethanolamine-N-methyltransferase was decreased. The addition of aluminum to the suboptimal zinc diet restored the activity of both enzymes to within the range of control values.

As regards intermediary metabolism, few effects of aluminum have been documented to date. Trapp has presented data that suggest that the aluminum ion is a dead-end inhibitor of yeast hexokinase (191). Also, aluminum inhibits the ferroxidase activity of ceruloplasmin (78).

That the neurofibrillary accumulations might result from a defect in axonal transport of filaments has been considered by several investigators (95,104,192). In the study by Liwnicz and others (104), the rate of fast axonal transport was similar in aluminum-treated and control animals. However, only fast transport was studied in this report, and neurofilaments are transported in the slow component (75). The finding of filamentous accumulations in the proximal axon (192) suggests an analogy to IDPN intoxication, in which a selective impairment of transport of the neurofilament triplet proteins has been demonstrated (63). An increased accumulation of choline acetyltransferase (ChAT) proximal to a sciatic nerve ligature in aluminum-treated rabbits (95) also suggests an abnormality in axonal transport. In the latter studies, we developed a model that employs direct micro-injection of concentrated aluminum chloride into the lumbar spinal cord, allowing prolonged survival of animals with neurofilamentous accumulations in a very high percentage of anterior horn cells giving rise to sciatic nerve axons. Because the development of neurofibrillary lesions in human brain is associated with a marked decline in choline acetyltransferase, we studied the quantity and transport of this enzyme in the lumber intraspinal aluminum model.

The rate of accumulation of choline acetyltransferase activity, referred to as the apparent rate of transport (139), was 4.1 ± 0.78 mm per 24 hours ($\bar{x} \pm$ S.E.M.) for the aluminum-treated rabbits and 1.9 ± 0.36 mm per 24 hours ($t = 2.55$, $df = 12$, $p < 0.02$) for the control animals. Also, the absolute choline acetyltransferase activity in unligated sciatic nerve was significantly less in the aluminum-treated animals (17.7 ± 1.9 nM acetylcholine per hour per 3-mm segment) than in controls (45.7 ± 2.3). The small amount of axonal degeneration noted in the intoxicated rabbits could not account for the decrement in ChAT activity (95).

Hetnarski and others (73) found that choline acetyltransferase and acetylcholinesterase (ACE) activities within the parenchyma of the spinal cords of aluminum-treated rabbits did not differ from controls. Yates and others (212), in contrast, found that ChAT but not acetylcholinesterase was significantly decreased in the gray matter of spinal cords showing tangles in about 90% of anterior horn cells. This study also compared the hypoglossal and vagal nuclei for enzyme activities because the former is severely affected with tangles and the latter is not. Both ChAT and ACE were decreased in the hypoglossal but not in the vagal nucleus. In all cases where the cholinergic enzymes were decreased in Yates's study, the enzyme glutamic acid dehydrogenase was not altered, suggesting a certain selectivity of the deficit. No ChAT loss was seen in areas generally not affected by aluminum-induced neurofibrillary lesions; for example, no change occurred in either frontal cortex or caudate nucleus, both of which are rich in cholinergic enzymes. Finally, a decrease in acetylcholinesterase has also been reported in cultured neuroblastoma cells exposed to aluminum salts (112).

Metallic Ions Other Than Aluminum

Although aluminum is preeminent as an agent inducing massive accumulation of perikaryal neurofilaments, toxic levels of other metals may also exert some effect on the neuron's filamentous skeleton. Intoxication of cultured rat dorsal root ganglion cells with cadmium ions leads to a striking perikaryal and axonal accumulation of 10-nm filaments (190); however, this ion causes no increase in neurofilaments in the same cells *in vivo* (152). O'Shea and Kaufman (119) applied copper sulfate to the neuroepithelium of embryonic mice and found occasional deposits of 10-nm filamentous material in the central regions of some cells. These cells were also depleted of microtubules; this fact lends support to the generalization that agents affecting microtubules may induce filamentous deposits. Lead has been postulated to bear some relationship to neurofibrillary degeneration. Niklowitz (116) administered tetraethyllead to rabbits and, after a latency of 6 to 12 hours, found bundles of straight 20-nm tubules in many cortical and hippocampal neurons. However, the nature of these structures and their relationship to neuronal intermediate filaments remains unclear. Other studies with lead in experimental animals have not reproduced this finding. Other metals that have been tested and do not appear to induce neurofibrillary changes include vanadium, indium, germanium, chromium, and manganese (206).

Hexacarbon Solvents

Certain industrial organic solvents containing aliphatic hydrocarbon backbones have been found to cause striking neurofilamentous lesions in humans and in experimental animals (1,41,72,81,94,110,145,165, 178,210). Intoxications with n-hexane, methyl n-butyl ketone (MBK), and their metabolites produce an initially distal and progressively centripetal axonal swelling and degeneration in selected pathways of the peripheral and central nervous systems. The subject has been extensively reviewed recently by Spencer and associates (179). The toxic hexacarbons are metabolized in part to 2,5-hexanedione, which is the compound primarily responsible for the neuropathic effects of the solvents. Studies on the metabolism of n-hexane and methyl n-butyl ketone demonstrated six intermediate metabolites that have a gradient of neurotoxic potency correlating with a peak serum level of the common metabolite 2,5-hexanedione (96,172). In hexacarbon neuropathies, giant axonal swellings first develop at the proximal portions of nodes of Ranvier in the distal, nonterminal regions of large myelinated fibers (175) (Figure 6-4). Later, swellings develop at internodal sites, and smaller myelinated and unmyelinated fibers

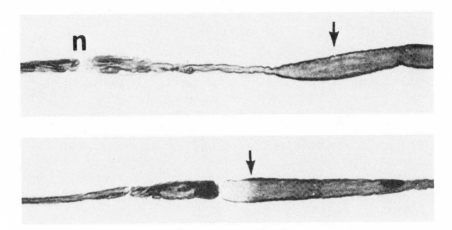

Figure 6-4. Single myelinated fibers teased from sciatic nerve (midthigh) of rat exposed intermittently to 1,300-ppm MBK for 3 months. (Top) Internodal fiber swelling (arrow) and relatively intact node of Ranvier (n) (×250). (Bottom) Fiber with paranodal swelling (arrow) and secondary retraction of myelin sheath (×250). (Fixed with buffered glutaraldehyde and osmium tetroxide, teased in epoxy resin.) (From reference 178. Used with permission.)

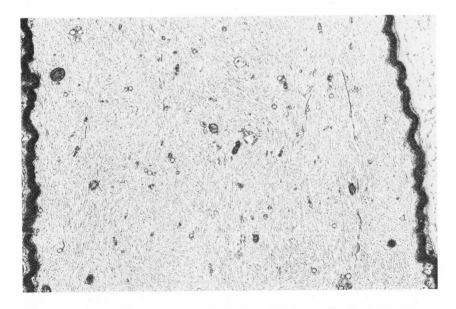

Figure 6-5. Swollen paranodal axon containing large numbers of neurofilaments and sparse neurotubules (×12,000). Longitudinal section of tibial nerve (1 cm above ankle) of rat exposed intermittently to 1300-ppm MBK for 4 months. (From reference 178. Used with permission.)

become affected. In the central nervous system, rostral regions of long ascending tracts and the caudal ends of long descending tracts are particularly vulnerable (176). Ultrastructurally, in addition to masses of 10-nm filaments, one sometimes sees clusters of mitochondria, microtubules, smooth endoplasmic reticulum, dense granules, and vesicles at the site of the axonal lesion (175) (Figure 6-5). Alterations of axonal transport have been documented in these structurally altered regions (67,109). Specifically, fast axonal transport is impaired, and it has been suggested that the filamentous accumulations block the passage of fast component material. Following injection of [^{3}H]-leucine into the spinal anterior horns of intoxicated animals, the region of distal axonal swelling shows increased grain densities by electron-microscopic autoradiography. The question of whether the alteration of transport is a direct effect of the toxin or is a consequence of neurofilamentous accumulation has recently been addressed by Griffin and associates (61). By directly applying 2,5-hexanedione to rat nerves *in situ*, they observed the earliest filamentous accumulation within 2 hours of exposure. Fast axonal transport studied at 6 hours, when the morphological effects of the toxin are well under way, was not yet blocked. The authors, therefore, suggested that the defect in fast transport occurs secondary to the cytoskeletal disorganization.

Whether the essential target of the hexacarbon solvents is the fibrous cytoskeleton remains unanswered. Unlike such filament-aggregating agents as colchicine and vinblastine, n-hexane and 2,5-hexanedione do not produce an increase in the number of 10-nm filaments nor do they alter microtubules in cultured C1300 murine neuroblastoma cells (159). This cell line has been shown to respond to aluminum salts by forming bundles of intermediate filaments (112). No similar filamentous change was noted after hexacarbon, although other, nonfilamentous cytopathological effects occurred. However, the principal cytopathological action of the hexacarbon solvents *in vivo* is seen in the distal portions of long axons; the lack of filamentous change in cultured neuroblastoma cells may relate to their short neuritic processes. In addition, untreated neuroblastoma cells appear not to contain detectable amounts of neurofilament triplet proteins, as assessed by gel electrophoresis. Recently, it has been reported that cultured skin fibroblasts exposed to 2,5-hexanedione display effects on intermediate filaments similar to those produced by colchicine (44). The filamentous aggregates induced by 2,5-hexanedione stained specifically with a monoclonal antibody to vimentin (44). Since vimentin appears to be the major protein constituent of neuroblastoma intermediate filaments, further studies are required to clarify the apparent differences in the response of vimentin filaments to 2,5-hexanedione in these two cell types. In an organotype spinal cord/nerve/muscle tissue culture system, Spencer and associates (177) and Veronesi and others (197) were able to reproduce precisely a hexacarbon distal axonopathy *in vitro*.

The mechanism of hexacarbon toxicity to neurofilaments has not been established. It has been proposed by Spencer and associates (171) that the hexacarbon solvents bind to and inhibit glycolytic enzymes along the nerve fiber. Indeed, the activity of glyceraldehyde-3-phosphate dehydrogenase and phosphofructokinase *in vitro* are inhibited by these compounds (144). An alternative hypothesis suggests that hexacarbons affect the function of neurofilaments by binding directly to them (149), and their neurotoxic properties are related to the γ-spacing of the carbonyl groups (38). This postulate has been expanded by Graham (57) on the basis of studies in which he incubated various proteins with the nonneurotic ketone, acetone or its neurotoxic dimer, 2,5-hexanedione. Only the latter incubations led to the formation of an orange chromophore, which is thought to occur concomitantly with the development of certain protein-protein cross-links. Progressive cross-linking of neurofilaments during slow transport could result in their aggregation and inability to pass through points of constriction proximal to the nodes of Ranvier, where they accumulate. Graham postulated a cross-link from the γ-diketone metabolic product of n-hexane to the ϵ-amino groups (e.g., lysine) of neurofilament proteins on the basis of the following evidence: (1) identical neurofilamentous lesions occur with other γ-diketones such as 2,5-heptanedione and 3,6-octanedione but not with the 2,3-, 2,4-, or 2,6-diketones (117,173,174); (2) incubation of 2,5-hexanedione with poly-L-lysine produced the same chromophore as occurred after incubation with various proteins (57).

Unlike the interspecies variability noted in the aluminum model, the hexacarbon solvents have produced axonal damage in all species tested (170). The filamentous neuropathy induced by the latter agents has been considered an experimental counterpart of the human axonopathies, giant axonal neuropathy (5) and infantile neuroaxonal dystrophy (107). These rare, chronically progressive polyneuropathies are also characterized by distal, focal axonal swellings containing chiefly 10-nm neurofilaments.

IDPN

The observation that β'-iminodipropionitrile (IDPN) induces neurofilamentous accumulation primarily in the proximal axon was initially made by Chou and Hartman (17,18) and shortly thereafter by Shimono and others (164). IDPN is a synthetic compound related to the toxic constituents of certain lathyrus plants known to cause diseases of both neural and connective tissue in humans and other animals. Lathyrism, a paralytic neurological disorder particularly affecting the lower extremities, was described by Hippocrates and was known even earlier in ancient India (8,160). When lathyrus peas were used instead of wheat to prepare flour following a poor wheat harvest, outbreaks of the disease occurred.

The intraperitoneal injection of IDPN induces a distinct behavior syndrome characterized by hyperkinesis in many different species (53,160). The associated neuropathologic lesions are largely localized to the spinal anterior horns, the reticular formation of brainstem, the area around the red nuclei, and the nuclei of various cranial nerves 18,167). Whether administered acutely by a single intraperitoneal injection or chronically in drinking water, IDPN induces fusiform enlargements of proximal axons (22). The duration of administration of the toxin alters the temporal pace at which these changes develop. The resultant fusiform swellings or giant axonal balloons, which measure more than 100 μ in diameter, are filled with 10-nm filaments (18,69,194,195). Although the individual filaments appear identical to normal neurofilaments, they are randomly oriented in abnormal fascicular arrays. Mitochondria within the axonal balloons often appear vacuolated. The relatively small amount of demyelination of peripheral nerves associated with IDPN intoxication is felt to be secondary to the axonal pathology. Clark and others (22) have found a reduction in axonal circumferences and cross-sectional areas in the distal axons of myelinated sciatic nerve fibers. This atrophy could not be accounted for by a selective loss of large fibers as assessed by histograms and was postulated by the authors to occur secondary to a defect in axonal transport. More recently, axonal swellings within the distal branches of the sciatic nerve in cats intoxicated with IDPN have been reported (62); in contrast to IDPN-treated rodents, the cats also displayed some degree of Wallerian degeneration. IDPN administered either intraperitoneally or in drinking water has also been found to affect the optic nerve as evidenced by swelling of the optic disc (123). With discontinuation of IDPN, these axonal swellings gradually disappear (123).

Various investigations of the effects of IDPN on neuronal metabolism have been carried out. An autoradiographic study of spinal motoneurons concluded that protein turnover was not appreciably altered in the perikarya of intoxicated rats, nor was there any significant *in situ* protein synthesis within the axonal balloons (19). Labeling with [3H] uridine in IDPN-treated mice, however, demonstrated increased RNA synthesis in nuclei of spinal ganglia after 1 hour and greater disappearance of the label from the cytoplasm after 24 hours compared to control mice (91). Enzymatic alterations in IDPN intoxication have also been examined. Schor and associates (154) studied a number of enzymes of carbohydrate metabolism in IDPN-treated rats and found only an increase in the activity of glucose-6-phosphate dehydrogenase. The authors interpreted the finding as nonspecific since dehydration and nerve transsection altered enzyme activity in a similar fashion. Cytochrome oxidase activity in brain from intoxicated rats is unchanged (11), as are the activities of the acid hydrolases (111). Gabay and Huang (50) reported a marked decrease in the specific activity of L-5-hydroxytryptophan:2-oxoglutarate aminotransferase but no effect on the other aminotransferases they studied. This report was followed up by Langlais and others (100), who found that serotonin levels were more markedly decreased than other amine neurotransmitters by IDPN administration and suggested a possible vulnerability of serotonergic neurons to the effects of IDPN.

The finding by Griffin and associates (63) that the proximal filamentous accumulations and distal loss of axonal area are associated with an abnormality in the slow component of axonal transport made IDPN intoxication the first filamentous model in which altered axonal transport could be directly implicated in the pathogenesis of the axonal degeneration. Yokoyama and associates (214) subsequently confirmed this conclusion. IDPN selectively impairs the slow component of axonal transport, which conveys neurofilaments; neither fast nor retrograde transport was impaired. In these experiments, radionuclides were injected into the rat lumbar ventral horn 1 to 2 days before and after IDPN treatment. Gel fluorograms of homogenized sciatic nerve segments were then performed (Figure 6-6). The slow transport peak failed to migrate beyond the initial 5 to 10 mm of the ventral roots. The neurofilament triplet polypeptides in particular did not pass beyond the most proximal segment. Tubulin and actin, also components of the slow transport system, similarly showed decreased transport. Griffin and others (63) also conducted experiments in which labeled slow-component proteins were allowed to pass well down the axon before IDPN was administered. Again, a large proportion of the neurofilament triplet proteins and substantial amounts of tubulin and actin were not further transported after toxin exposure (63).

It is currently postulated that this defect in slow axonal transport may be preceded by a reorganization of the axonal cytoskeleton (61, 120). IDPN given systemically or injected directly into rat sciatic nerve results in a displacement of neurofilaments toward the periphery of the axon. The peripheral ring of chaotically oriented neurofilaments surrounds a central channel containing normal appearing

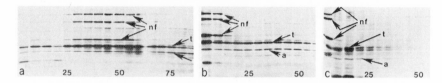

Figure 6-6. Fluorograms of polyacrylamide slab gels, showing the distribution of individual labeled proteins along the nerves. Each tack represents a 5-mm nerve segment, with the most proximal segment at the far left; the numbers (25, 50, or 75 mm) identify the distance of the segments along the nerve. (a) Control animal 21 days after labeling with [^{35}S] methionine; 5% to 17.5% gradient gel; exposure, 6 weeks. The neurofilament triplet proteins (nf) (68,000, 145,000, and 200,000 mol wt) are present primarily in tracks corresponding to 25 to 50 mm (that is, they have moved 1.0 to 2.5 mm per day). Tubulin (t) and actin (a) are present over a wider spectrum of velocities (0.5 to 5.0 mm per day). (b) Experimental animal 21 days after intraspinal injection of [^{35}S] methionine and 20 days after injection of IDPN; 5% to 17.5% gel; exposure, 6 weeks. The neurofilament triplet proteins are in large part retained in the initial two segments; transport of tubulin and actin is impaired as well, compared to control. (c) Experimental animal given IDPN in drinking water, 21 days after labeling with [^{3}H] leucine; 7.5% gel; exposure, 4 months. (In the first track an artifact in the X-ray film is present between the tubulin and actin bands.) The marked impairment in slow transport is apparent. (From reference 63. Used with permission.)

microtubules, smooth endoplasmic reticulum, and mitochondria. This change occurs simultaneously along the entire length of the sciatic nerve. It has, therefore, been suggested that an IDPN-induced impairment of the interactions between neurofilaments and microtubules may underlie the defect in slow transport (120). Furthermore, these findings imply that neurofilaments are not essential for normal fast transport, which according to electron-microscopic autoradiography occurs normally but is confined to the central channel of the axon in IDPN-treated rats (121). A similar structural reorganization of the cytoskeleton appears to occur in the hexacarbon neuropathies (61, 120) and in chronic constriction of nerves (120).

Mitotic Spindle Inhibitors

The finding that colchicine exerts its properties of mitotic spindle inhibition by binding to microtubules was reported by Taylor in 1965 (187). Because important interactions between microtubules and neurofilaments have been postulated (137,143,162), the effect of mitotic spindle inhibitors on intermediate filaments has also been extensively studied. Whether mitotic spindle inhibitors act directly on intermediate filaments or through their effects on microtubules remains unclear. The effects of colchicine on intermediate filaments have been best demonstrated in various nonneural cells in culture. Interphase HeLa cells exposed to colchicine produce large numbers of filaments (141). Addition of colchicine to fibroblasts causes their intermediate filaments to aggregate into perinuclear whorls (54,76, 82). Similarly, antimitotic agents induce juxtanuclear filamentous caps in baby hamster kidney cells (181). Staining with antibodies against vimentin, the 58,000-mol wt component of these 10-nm filaments, immunolabels the aggregated filaments (79).

The aggregation of neurofilaments by colchicine has been demonstrated in cultured neurons (131,156). By exposing neonatal mouse dorsal root ganglion cultures to the spindle inhibitor maytanprine, Sato and others (148) reported an 80% reduction in the microtubule population and a doubling in the number of neurofilaments. These workers suggested that the accumulation of filamentous tangles did not occur via the synthesis of new neurofilament proteins but by the rearrangement and retrograde transport of existing neurofilament proteins. In partial support of this hypothesis is the finding that total protein synthesis measured by [^{14}C]-leucine incorporation into proteins is not increased when chick dissociated spinal ganglia are treated with colchicine or colcemid (33). Because of the disruption of microtubules, neurite outgrowth was inhibited in the latter cultures.

In the central nervous system of rabbits, the intracisternal injection of colchicine or other mitotic spindle inhibitors such as vinblastine, vincristine, or podophyllotoxin induces an accumulation of filaments in the cytoplasm of the perikarya, axons, and dendrites (153, 163,204,207). In such neurons, the rough endoplasmic reticulum appears to clump, and there is a loss of microtubules. These effects of spindle inhibitors appear to be spontaneously reversed after approximately 2 weeks in those animals that survive the neurological dysfunction produced by the agents (204). Colchicine differs from vinblastine and vincristine in that it does not induce the paracrystalline inclusions seen with the latter agents (74,153). By electron-microscopic immunolocalization, the paracrystalline inclusions stain with an antibody against the 210,000-mol wt microtubule-associated protein (37). Recently, another agent, doxorubicin, which impairs DNA-dependent RNA synthesis, has been found to induce neurofilamentous accumulations in the perikaryon and proximal axon when injected into the eyes of rats or guinea pigs (122).

The great variability in susceptibility of neuronal populations to mitotic spindle inhibitors has been studied by Goldschmidt and Steward (55,56). They injected colchicine directly into the dentate gyrus in adult rats; this resulted in extensive destruction of dentate granule cells while hippocampal pyramidal cells were spared (55). Cerebellar injections caused destruction of both granule cells and Purkinje cells while sparing neurons in the molecular layers (56). After striatal injection, neuronal damage occurs (55); however, it is less extensive than in hippocampus and cerebellum (56).

Recently, maytansine has been added to the list of antimitotic agents that bind to tubulin (106,140) and can induce neurofilamentous accumulations (52). Maytansine is an ansa macrolide obtained from the African plant Maytenus. The neurons of rabbits injected intrathecally with maytansine show an early decrease in microtubules accompanied by both dendritic and perikaryal accumulation of 10-nm filaments morphologically indistinguishable from normal neurofilaments (52). In addition, the rough endoplasmic reticulum appears to aggregate adjacent to the nucleus. Using dissociated adult mouse dorsal root ganglia, Kim and others (89) have induced neurofibrillary changes by exposure to maytansine. When maytansine was removed from the media, the cells reverted to normal morphology within 3 to 4 days. Analogous to the effect of colchicine on fibroblasts, maytansine induces in Schwann cells the aggregation of intermediate filaments, which, by immunofluorescent staining, are presumed to be vimentin filaments (89).

Because of their effect on microtubules, all of the mitotic spindle inhibitors, particularly colchicine, have been used to study axonal transport. Under various experimental conditions, these agents have been shown to interrupt fast transport (97,124,125,146). However, the relation of an axonal transport abnormality to the neurofilamentous lesions remains unresolved (10,59).

Acrylamide

The effects of acrylamide administration on the central nervous system were first studied by Kuperman (98) and later by McCollister and others (105). In 1966, Fullerton and Barnes (49) identified the peripheral nervous system as the primary target of acrylamide intoxication. Shortly thereafter, Prineas (136) reported the ultrastructural hallmark of acrylamide neuropathy — a marked increase in the number of neurofilaments in the terminal axoplasmic expansions of affected nerves. In addition, the mitochondria within the axoplasm were enlarged and showed an increased number of cristae, the axonal plasma membranes displayed a complex unfolding, and microtubules were relatively decreased in number. Similar though less marked changes were demonstrated in axons of the central nervous system. Neuronal perikarya were not found to contain accumulations of neurofilaments. More recently, however, some perikaryal alterations have been reported in acrylamide neuropathy (182). Rats exposed to acrylamide monomer developed in the cell bodies of dorsal root ganglia a cytoplasmic remodeling, with mitochondria, Golgi complexes, and free polyribosomes forming a dense perinuclear inner zone and the rough endosplasmic reticulum forming an outer zone.

Several investigators have examined biochemical changes accompanying acrylamide neurotoxicity. Certain analogues of acrylamide such as N-hydroxymethylacrylamide, N-methylacrylamide and N,N'-diethylacrylamide are also capable of inducing the neuropathy, although they are considerably less effective than the parent compound (45). Using [^{14}C]-labeled acrylamide, Hashimoto and Aldrige (70) demonstrated approximately equal distribution of radioactivity in various brain subcellular fractions. Kaplan and Murphy have found that β-glucuronidase activity is increased in nerves affected by acrylamide (85) and that both DDT and phenobarbitol partially protect rats from developing the neuropathy (86). Morphologically, there is a resemblance between acrylamide neuropathy and the neuropathy seen in vitamin E-deficient animals; both are characterized by an accumulation of neurofilaments in the axon terminus.

Because colchicine is known to bind specifically to tubulin and may exert its effect on neurofilaments as a result of this interaction, Rasool and Bradley (139) studied the colchicine-binding properties of tubulin in rats previously exposed to acrylamide. A 75% decrease in tubulin-colchicine binding was observed in animals with clinical neuropathies induced by acrylamide. It is generally agreed that there is no alteration in the rate of slow transport and only a minimal reduction in the rate of fast transport in acrylamide neuropathy (183); the latter correlated poorly with the extent of histological change in the nerve (139). However, although the overall rates of axoplasmic transport systems may not be greatly altered in this distal axonopathy, it is possible that interference with transport of specific molecules has occurred. In this regard, alterations in the axonal transport of acetylcholinesterase and choline acetyltransferase have been reported (138). Acrylamide-treated rats had a decreased acetylcholinesterase transport rate in sciatic nerve in both single- and double-ligature experiments. There was an increase in the size of the mobile fraction of acetylcholinesterase and a decrease in the total amount of enzyme transported. The transport rate of choline acetyltransferase did not change, but the amount of this enzyme transported actually increased. One of the problems with such ligature studies, which are often utilized in the study of enzymatic transport, is the possibility of secondary effects produced by local injury to the axon. Cavanagh and Gysbers (16) found a "dying-back" degeneration proximal to the level of a ligature around the posterior tibial nerve in intoxicated rats. The severity of this degeneration does not appear to bear any relationship to the distance of the nerve injury from the soma (161). Griffin and others (64,66) have demonstrated a failure of regeneration after crush injury in acrylamide-intoxicated rats. Labeled protein was found in the terminal axonal expansions; this implies that transport through the injured site occurred but that the transported material was not properly utilized for regeneration. Souyri and others (169) observed local stasis of fast-transported proteins in the distal axon that suggests that the axoplasmic filamentous swellings may quantitatively reduce the amount of transported material able to pass through them. These workers studied nerves early in the course of acrylamide neuropathy. They hypothesize that local disorganization of peripheral smooth endoplasmic reticulum, one of the earliest detectable changes in the axons of acrylamide-treated animals, may be responsible for the focal stasis of fast-transported material (21).

Other Neurotoxins

Carbon Disulfide

A number of other chemical agents known to produce filamentous pathology have not been extensively utilized for experimental studies of neurofilaments. Polyneuropathy following prolonged exposure of rats and rabbits to carbon disulfide is characterized by increased quantities of neurofilaments and an associated disappearance of microtubules within peripheral axons (84,102,186). The filamentous accumulations are seen as numerous paranodal and internodal fusiform swellings along the axon, often associated with Wallerian degeneration distal to the swelling (180). Rat spinal cord neurofilaments isolated after intraperitoneal injection of carbon disulfide showed extensive binding of the toxin (150). Free thiol groups in the neurofilament proteins are believed to bind with the agent (107). The human filamentous neuropathy associated with prolonged use of disulfiram (Antabuse) (2,113) may occur as a result of metabolism of the drug through disulfide intermediates (77). Disulfiram produces an analogous experimental distal neuropathy in rats (3).

Buckthorn Neuropathy

Buckthorn neuropathy, which had previously been considered a demyelinating process, has recently been shown in organotypic cultures of mouse spinal cord and dorsal root ganglia to involve primarily the axon, including, among other changes, filamentous lesions (71). The buckthorn shrub (*Karwinskia humboldtiana*) produces a fruit from which four related neurotoxic compounds have been isolated (40). The primary neuronal toxicity of these agents was first suggested by Aoki and Muñoz-Martinez (4). There is a redistribution of axonal organelles in a pattern opposite to that seen with IDPN and n-hexane. The toxins isolated from the buckthorn shrub produce a central accumulation of neurofilaments and a peripheral displacement of microtubules and smooth endoplasmic reticulum. During the first 2 weeks of toxic exposure, there is a paucity of other organelles within the central channel. After 2 weeks, the additional findings of stacks of smooth endoplasmic reticulum appears to be in association with the microtubules in the periphery. A primary or secondary defect in fast axonal transport has been proposed in view of the disorganization of the smooth endoplasmic reticulum channels (71).

Tri-ortho-cresyl Phosphate

The neurotoxic organophosphorous compound tri-ortho-cresyl phosphate has been extensively studied in experimental animals. In exposed animals, neurofilaments first aggregate and then are replaced by electron-dense granular material (9,135). There is no accumulation of neurofilaments. However, microtubules and membrane-bound vesicles accumulate in the distal axon.

Ethanol

An intriguing observation that requires further investigation is the finding of pairs of helically wound 10-nm filaments with a periodicity of 35 nm in young rats given ethanol for at least 6 months (199). Such filaments, although morphologically distinguishable, do bear a resemblance to the paired helical filaments seen in Alzheimer's disease and other human neurofibrillary degenerations.

Axotomy

Axotomy induces alterations in neurofilament organization during the course of Wallerian degeneration. By 72 hours after nerve section, the filaments fragment, aggregate, and intertwine (15). Particularly in large myelinated fibers, some accumulation of neurofilaments and loss of their lengthwise orientation is apparent as early as 19 hours after axotomy (101). As in the mammalian studies noted above, section of the avian optic nerve results in the appearance of neurofilaments in neurons of the contralateral optic tectum (58). Just as in certain toxic neuropathies, Wallerian degeneration induces the focal accumulation of axonal mitochondria (200). Soifer and others (168) studied the protein composition of nerves undergoing Wallerian degeneration and, as predicted from ultrastructural findings, documented a loss of the neurofilament triplet proteins. A 65,000-mol wt protein associated with neither the neurofilaments nor myelin also disappeared.

Vitamin-E Deficiency

At least one deficiency syndrome has been specifically associated with neurofilamentous accumulations. In rats made deficient in vitamin E, an increased number of neurofilaments and a swelling of the axon has been reported to occur in the terminal portion of the peripheral nerve axon (99,128). Swollen dystrophic axons or spheroids were prominent in major sensory nuclei of the caudal medulla, fasciculus gracilis, fasiculus cuneatus, and posterior horns. A number of other ultrastructural abnormalities were also noted in this study, including focal collections of mitochondria and numerous electron-dense bodies. Similar changes have been reported in human peripheral nerve-biopsy specimens from patients who may have vitamin-E deficiences secondary to gastrointestinal malabsorption (24,51,184, 185). Rhesus monkeys, however, may not be as susceptible as humans to the effects of vitamin-E deficiency since axonal swellings were only rarely seen in this species by Nelson and associates (115). These authors did note distal large-fiber axonal degeneration and a predisposition of sensory tracts to the lesions. Wood and Boegman (209) did not find any alteration in slow-axonal transport in vitamin E-deficient rats, unlike findings with IDPN. Fast transport, however, was increased 2.4-fold above control levels.

Temperature Effects on Intermediate Filaments *in Situ*

It has long been known that an extensive network of neurofibrils can be seen light microscopically in the neurons of lizards subjected to low temperatures (12,188). These changes have more recently been confirmed electron microscopically by Potter and associates (133, 134). Roots and Bordar (142) have demonstrated a similar phenomenon in goldfish, in which neurofilaments are seen in the synaptic boutons of the brain only in fish maintained at 5°C for 177 days or more. Studies in cultured fibroblasts that examined the effects of cold treatment on cytoskeletal components showed no effect on the intermediate filaments but a disappearance of microtubules (198). Hence, under these conditions, normal intermediate filament organization is maintained independent of microtubules.

Naturally Occurring Animal Models
of Neurofibrillary Degeneration

Numerous human neurological diseases are characterized by idiopathic, progressive attrition of selected neuronal populations. Genetic factors are known to underlie the neuronal abiotrophy in some of these diseases and may well play some role in virtually all such disorders. Consequently, the existence of naturally occurring models of specific neuronal degenerations in inbred animals provides a major resource for detailed pathogenetic studies. In recent years, several neuronal and/or axonal degenerations in mammals have been described that are characterized by prominent neurofilamentous pathology. The major features of some of these models will now be reviewed.

Hereditary Canine Spinal Muscular Atrophy

Proximal axonal swellings stuffed with 10-nm filaments have been reported in sporadic human motor neuron disease (amyotrophic lateral sclerosis) (14,20). There is evidence that these neurofilamentous lesions may represent an early pathological change. Patients who die early in the course of the disease display axonal swellings in spinal segments that showed active involvement clinically (14,20). It should be noted, however, that such filamentous swellings can be found in spinal cords of neurologically normal humans (23). Recently, Cork and associates described a family of purebred Brittany spaniels with progressive motor neuron degeneration in the spinal cord and brainstem (26). Many of the surviving motor neurons had filamentous swellings in proximal axons highly similar to those described in human amyotrophic lateral sclerosis. Great numbers of swellings were observed in young animals with an accelerated form of the disease (25), which raised the possibility of early reorganization or failed transport of neurofilaments in the canine degeneration. In subsequent studies, Griffin, Cork, and associates have found an impairment of slow transport of neurofilament proteins, tubulin, and actin in these dogs (60). Preliminary results suggest greater abnormality of slow transport in young dogs with the accelerated disease. As might be expected from the transport data, the axonal changes in the Brittany spaniels have morphological similarity to IDPN-induced lesions. In both models, some neurofilaments accumulate in the perikaryon, but the major pathology is found in proximal internodes of the axon, where skeins of maloriented neurofilaments surround a central channel of microtubules (27). Secondary demyelination occurs around the abnormal regions of the axons. Loss of anterior horn cell bodies and Wallerian degeneration of axons is seen in the canine model but not in IDPN-treated animals. The disorder in this small family of inbred Brittany spaniels appears to be inherited in an autosomal recessive manner (26).

Similar axonal swellings are occasionally seen in the spinal ventral horns of the wobbler mouse (Mitsumoto and Bradley, personal communication), another inherited model of motor neuron degeneration. The presence of neurofilament accumulations within the perikarya of anterior horn cells has also been described in a cat dying of sporadic motor neuron disease (196).

Canine Giant Axonal Neuropathy

A rare, recessively inherited axonal degeneration that expresses itself in childhood as a progressive peripheral neuropathy usually associated with mental retardation and abnormal, tightly curled hair has been described in humans (5,126). This entity, called giant axonal neuropathy, has as its principal pathological feature the presence of multifocal filamentous swellings in the distal portions of both peripheral and central axons, a pattern similar to that seen in hexacarbon neuropathy. There is evidence that this disease may also involve filaments in glial, Schwann, and endothelial cells, which suggests a generalized reorganization of filamentous cytoskeletons (126). Duncan and Griffiths have found a naturally occurring central-peripheral distal axonopathy with this form of pathology in an inbred Alsatian dog (42). Tightly packed whorls of neurofilaments were found in distal nonmyelinated and myelinated peripheral fibers in the latter case associated with attenuation of the myelin sheath (43). In some enlarged axons, there was a notable separation of microtubules, mitochondria, and smooth endoplasmic reticulum from the mass of neurofilaments. Some Schwann cells appeared to participate in the filamentous reorganization by displaying accumulations of microfilaments in their cytoplasm. In the central nervous system, axonal spheroids containing disorganized neurofilaments were present in the distal portions of long ascending and descending tracts of spinal cord and brainstem (68). Similar axonal lesions were found diffusely in the cerebral cortex. The pathology is similar to that of human giant axonal neuropathy. Neurofilament-enriched fractions prepared from central nervous system and sciatic nerves of affected dogs showed the presence of neurofilament polypeptides at 68,000, 145,000 and 230,000-mol wt, with a relative distribution indistinguishable from that of normal dogs (83). Studies of axonal transport of intermediate filaments in the canine model have not been reported at this time.

Troncoso and associates very recently identified a recessively inherited clinicopathological syndrome in purebred Rottweiler dogs characterized by progressive ataxia of gait and the presence of distal axonal swellings in the dorsal and ventral horns of the spinal cord, the gracilis, cuneatus and trigeminal nuclei, and the cerebellar cortex and deep nuclei (193). The authors propose this as a model of the human disorders referred to as infantile and late-onset neuroaxonal dystrophy. The central nervous system changes in this entity also resemble those in canine giant axonal neuropathy.

Swayback in Sheep

It has been recognized for many years that young lambs may occasionally develop a progressive and sometimes fatal ataxia of gait, with clinical onset noted at birth or within a few weeks thereafter (7). The appearance of this neurological disorder in certain flocks has been associated with deficiencies of blood and hepatic copper levels, presumably related to inadequate dietary intake, although the precise role of low copper in the pathogenesis of the disease is not clear. The characteristic neuropathological changes comprise abnormalities of both nerve cell bodies and axons in large neurons of certain brainstem nuclei and in anterior horn cells of spinal cord (7). Ultrastructurally, the most prominent and consistent abnormality was a vast increase of cytoplasmic filaments morphologically indistinguishable from those encountered in adjacent normal neurons (13). Severely affected neurons showed swollen cell bodies filled almost exclusively with filaments. Axonal fibrillary changes were much less marked than those in the perikaryon. In most abnormal neurons, other altered organelles were found, including enlarged, granular mitochondria, abnormally organized Golgi apparatus, and a progressive loss of free and membrane-bound ribosomes.

Conclusion

An analysis of completed and ongoing studies on experimental models of filamentous pathology demonstrates the contribution of such investigations to our knowledge about the role of intermediate filaments both in normal neuronal processes and in certain disorders of the human central and peripheral nervous system. For example, recent experiments employing direct application of filament-reorganizing neurotoxins to peripheral nerves have provided new information indicating that fast axonal transport may continue following a fundamental rearrangement of neurofilaments and microtubules within axons and that such fast transport can occur in areas of the axoplasm devoid of neurofilaments. Future studies that expand on this work should provide mechanistic information about the structural requirements for a normally functioning axonal transport system that would not be available from investigations of normal animals. Similarly, use of appropriate animal and tissue culture models may provide new insights into the synthesis and degradation of neurofilament proteins during neuronal development and regeneration as well as the interactions of filaments with other cytoplasmic orangelles.

In the area of cytopathologic studies, the use of hexacarbon solvents, IDPN, acrylamide, and aluminum as experimental agents has yielded data and hypotheses relevant to the mechanisms of axonopathies not only in the corresponding human intoxications but also in spontaneously occurring degenerations such as Friedreich's ataxia and amyotrophical lateral sclerosis. As another example, the discovery of naturally occurring genetic models of motor neuron disease and giant axonal neuropathy in inbred dogs should enable a wealth of dynamic pathogenetic studies that are not possible in humans.

On the other hand, there remains a major class of human neuro-fibrillary diseases for which no adequate experimental model current-ly exists. The mechanism of progressive, age-related accumulation of abnormal intraneuronal fibers that appear to show some structural and antigenic relationship to normal neurofilaments remains a funda-mental unresolved problem in human neurobiology. In contrast to the numerous human disorders in which axons and perikarya accu-mulate filaments with structural and biochemical properties indistin-guishable from normal filaments, the paired helical filaments found in selected neurons during aging and in Alzheimer's disease have re-cently been shown to have very unusual chemical characteristics (157). Their uniqueness lies in their remarkable insolubility in a wide range of protein denaturants, reducing agents, and harsh solvents that readily bring the vast majority of intracellular proteins, including normal neurofilaments, into solution. Recent studies have demon-strated that paired helical filaments remain ultrastructurally intact after extensive treatment with sodium dodecyl sulfate, urea, guani-dine hydrochloride, formic acid, potassium thiocyanate, dilute hydrochloric acid, and various proteolytic enzymes (80,157). These findings indicate that paired helical filaments are high-molecular-weight polymers that are held together by covalent bonds other than disulfides (i.e., protein-protein cross links) and/or by extensive hy-drophobic forces and hydrogen bonding (See Chapter 7).

At present, aged neurons of subhuman primates and other mam-mals have not been observed to contain paired helical filaments. No animal or *in vitro* model of these filamentous structures is known to exist. Although certain susceptible strains of inbred mice have been shown to develop neuritic plaques in gray matter many months after brain inoculation of the scrapie agent (202), the abnormal neurites of these induced plaques do not contain paired helical filaments. The induction of abnormal fibrils consisting of apparent intermediate filaments twisted into helices has been described in cultured rat dor-sal root ganglia neurons exposed to ethanol (199). The conformation and periodicity of these structures is, however, different from those of human paired helical filaments. Naturally occurring helices of neu-rofilaments have been observed in neurons of the whip spider (48); again, their ultrastructure is distinct from that of Alzheimer paired helical filaments.

The molecular mechanisms leading to the assembly of paired heli-cal filaments in human neurons may well turn out to be highly com-plex, perhaps involving important posttranslational modifications of neurofilaments or other normal proteins, followed by subsequent enzymatic or spontaneous cross linking reactions. It may, therefore, be exceedingly difficult to devise experimental systems in animals or in cultured neurons that reproduce some or all of the features of the Alzheimer-type neurofibrillary degeneration. Nonetheless, efforts to develop such models, based on emerging information about the nature of the paired helical filaments themselves, must be pursued in the coming years if the role of altered intermediate filaments in age-related human neuronal degeneration is to be understood.

REFERENCES

1. Allen, N., J. R. Mendell, D. J. Billmaier, R. E. Fontaine, and J. O'Neill. 1975. Toxic polyneuropathy due to methyl n-butyl keton. *Arch. Neurol. 32*:209-218.

2. Ansbacher, L. E., E. P. Bosch and P. A. Cancilla. 1982. Disulfiram neuropathy. A neurofilamentous distal axonopathy. *Neurology (N.Y.) 32*:424-428.

3. Anzil, A. P., and S. Dozic. 1978. Disulfiram neuropathy: An experimental study in the rat. *J. Neuropath. Exp. Neurol. 37*:585 (abs.).

4. Aoki, K., and E. J. Muñoz-Martinez. 1981. Quantitative changes in myelin proteins in a peripheral neuropathy caused by Tullidora (*Karwinskia humboldtiana*). *J. Neurochem. 36*:1-8.

5. Asbury, A. K., M. K. Gale, S. C. Cox, J. R. Baringer, and B. O. Berg. 1972. Giant axonal neuropathy: a unique case with segmented neurofilamentous masses. *Acta Neuropathol. 20*:237-247.

6. Baes, C. F., and R. E. Mesmer. 1976. The Hydrolysis of Cations. Wiley, New York, p. 112.

7. Barlow, R. M., D. Purves, E. J. Butler, and I. J. Macintyre. 1960. Swayback in southeast Scotland: II. Clinical, pathological and biochemical aspects. *J. Comp. Path. 70*:411-427.

8. Barrow, M. W., C. F. Simpson and E. J. Miller. 1974. Lathyrism: A review. *Quart. Rev. Biol. 49*:101-128.

9. Bischoff, A. 1967. The ultrastructure of tri-ortho-cresyl phosphate poisoning: I. Studies on myelin and axonal alterations in the sciatic nerve. *Acta Neuropathol. (Berl.) 9*: 158-174.

10. Bradley, W. G., and M. H. Williams. 1973. Axoplasmic flow in axonal neuropathies: I. Axoplasmic flow in cats with toxic neuropathies. *Brain 96*:235-246.

11. Brownlow, E. K., and H. Heath. 1969. Biochemical changes in β, β'-iminodipropionitrile-treated rat brain and prevention of toxicity by ethionine. *J. Neurochem. 16*:567-575.

12. Cajal, S. R. Y. 1904. Variaciones morfologicas normales y patologicas. *Trab. Lab. Invest. Biol. 3*:9-15.

13. Cancilla, P. A., and R. M. Barlow. 1966. Structural changes of the central nervous system in swayback (enzootic ataxia) of lambs. *Acta Neuropathol. 6*:251-259.

14. Carpenter, S. 1968. Proximal axonal enlargement in motor neuron disease. *Neurology 18*:841-851.

15. Causey, G., and H. Hauffman. 1955. Axon sprouting in partially deneurotized nerves. *Brain 78*:661-668.

16. Cavanagh, J. B., and M. F. Gysbers. 1980. "Dying-back" above a nerve ligature produced by acrylamide. *Acta Neuropathol. (Berl.) 51*:169-177.

17. Chou, S. M., and H. A. Hartmann. 1964. Axonal lesions and waltzing syndrome after IDPN administration in rats. *Acta Neuropathol. 3*:428-450.

18. Chou, S. M., and H. A. Hartmann. 1965. Electron microscopy of focal neuroaxonal lesions produced by β, β′-iminodipropionitrile (IDPN) in rats. *Acta Neuropathol.* 4:590-603.

19. Chou, S. M., and R. A. Klein. 1972. Autoradiographic studies of protein turnover in motoneurons of IDPN-treated rats. *Acta Neuropathol. (Berl.)* 22:183-189.

20. Chou, S. M., J. D. Marton, J. A. Gutrecht, and H. G. Thompson. 1970. Axonal balloons in subacute motor neuron disease. *J. Neuropath. Exp. Neurol.* 29:141-142.

21. Chretien, M., G. Patey, F. Souyri, and B. Droz. 1981. "Acrylamide-induced" neuropathy and impairment of axonal transport of proteins: II. Abnormal accumulations of smooth endoplasmic reticulum as sites of focal retention of fast transported proteins: Electron microscope radioautographic study. *Brain Res.* 205:15-28.

22. Clark, A. W., J. W. Griffin, and D. L. Price. 1980. The axonal pathology in chronic IDPN intoxication. *J. Neuropath. Exp. Neurol.* 39:42-55.

23. Clark, A. W., I. M. Parhad, J. W. Griffin, and D. L. Price. 1982. Neurofilamentous axonal swellings in the lumbosacral spinal cord of normal man. *J. Neuropath. Exp. Neurol.* 41:379.

24. Cooke, W. T., A. G. Johnson, and A. L. Woolf. 1966. Vital staining with electron microscopy of the intramuscular nerve endings in the neuropathy of adult celiac disease. *Brain* 89:663-682.

25. Cork, L. C., J. W. Griffin, R. J. Adams, and D. L. Price. 1980. Animal model: Hereditary canine spinal muscular atrophy. *Am. J. Path.* 100:599-602.

26. Cork, L. C., J. W. Griffin, J. F. Munnell, M. D. Lorenz, R. J. Adams, and D. L. Price. 1979. Hereditary canine spinal muscular atrophy. *J. Neuropath. Exp. Neurol.* 38:209-221.

27. Cork, L. C., J. W. Griffin, and D. L. Price. 1982. A comparison of the neuropathology of hereditary canine spinal muscular atrophy and β, β′-iminodipropionitrile (IDPN). *J. Neuropath. Exp. Neurol.* 41:373 (abs.).

28. Crapper, D. R. 1973. Experimental neurofibrillary degeneration and altered electrical activity. *Electroenceph. Clin. Neurophysiol.* 35:575-588.

29. Crapper, D. R., and A. J. Dalton. 1973. Aluminum induced neurofibrillary degeneration, brain electrical activity and alterations in acquisition and retention. *Physiol. Behav.* 10:935-945.

30. Crapper, D. R., and G. J. Tomko. 1975. Neuronal correlates of an encephalopathy associated with aluminum neurofibrillary degeneration. *Brain Res.* 97:253-264.

31. Czosnek, H., D. Soifer, and Wisniewski, H. 1978. Protein biosynthesis in Al-induced neurofibrillary changes. *In*: VIII International Congress of Neuropathology (Abstracts). Washington, D.C., p. 604.

32. Dahl, D., and A. Bignami. 1978. Immunochemical cross-reactivity of normal neurofibrils and aluminum-induced neurofibrillary tangles: Immunofluorescence study with anti-neurofilament serum. *Exp. Neurol.* 58:74-80.

33. Daniels, M. P. 1972. Colchicine inhibition of nerve fiber formation *in vitro*. *J. Cell Biol.* 53:164-176.

34. DeBoni, U., A. Otvos, J. W. Scott, and D. R. Crapper, 1976. Neurofibrillary degeneration induced by systemic aluminum. *Acta Neuropath. (Berl.)* 35:285-294.

35. DeBoni, U., J. W. Scott, and D. R. Crapper. 1974. Intracellular aluminum binding: A histochemical study. *Histochemistry* 40:31-37.

36. DeBoni, U., M. Seger, and D. R. Crapper McLachlan. 1980. Functional consequences of chromatin-bound aluminum in cultured human cells. *In*: Aluminum Neurotoxicity. L. Liss, editor. Pathotox Publishers, Park Forest South, Ill., pp. 65-81.

37. DeBrabander, M., J. C. Bulinski, G. Guens, J. DeMey, and G. G. Borisy. 1981. Immunoelectron microscopic localization of the 210,000 mol wt microtubule-associated protein in cultured cells of primates. *J. Cell Biol.* 91:438-445.

38. DiVincenzo, G. D., C. J. Kaplan, and J. Dedinas. 1976. Characterization of the metabolites of methyl n-butyl ketone, methyl iso-butyl ketone, and methyl ethyl ketone in guinea pig serum and their clearance. *Toxicol. Appl. Pharmacol.* 36:511-522.

39. Döllken. 1898. Uber die Wirkung des Aluminum, mit besonderer Berücksichtigung der durch des Aluminum versursachten Läsionen im Zentralnervensystem. *Nauyn-Schmiederbergs Arch. exp. Path. Pharmak.* 40:98-120.

40. Dreyer, D. L., I. Arai, C. D. Bachman, W. R. Anderson, Jr., R. G. Smith, and G. D. Daves, Jr. 1975. Toxins causing non-inflammatory paralytic neuropathy: Isolation and structure elucidation. *J. Am. Chem. Soc.* 97:4985-4990.

41. Duckett, S., N. Williams, and S. Frances. 1974. Peripheral neuropathy associated with inhalation of methyl n-butyl ketone. *Experientia* 30:1283-1284.

42. Duncan, I.D., and I. R. Griffiths. 1977. Canine giant axonal neuropathy. *Vet. Rec.* 101:438-441.

43. Duncan, I. D., and I. R. Griffiths. 1979. Peripheral nervous system in a case of canine giant axonal neuropathy. *Neuropath. Appl. Neurobiol.* 5:25-38.

44. Durham, H. D., and S. D. J. Pena. 1982. The neurotoxin 2,5-hexanedione induces aggregation of intermediate filaments in cultured skin fibroblasts. *Neurology (N.Y.)* 32:A215 (abs.).

45. Edwards, P. M. 1975. Neurotoxicity of acrylamide and its analogues and effects of these analogues and other agents on acrylamide neuropathy. *Brit. J. Ind. Med* 32:31-38.

46. Embree, L. J., A. Hamberger, and J. Sjöstrand. 1967. Quantitative cytochemical studies and histochemistry in experimental neurofibrillary degeneration. *J. Neuropath. Exp. Neurol.* 26:427-436.

47. Exss, R. E., and G. K. Summer. 1973. Basic proteins in neurons containing fibrillary deposits. *Brain Res.* 49:151-164.

48. Foelix, R. F., and M. Hauser. 1979. Helically twisted filaments in giant neurons of a whip spider. *Eur. J. Cell Biol.* 19:303-306.

49. Fullerton, P. M., and J. M. Barnes. 1966. Peripheral neuropathy in rats produced by acrylamide. *Br. J. Ind. Med.* 23:210-221.

50. Gabay, S., and K. Huang. 1969. Studies of brain aminotransferase and psychotropic agents: I. Interaction of β-β'-iminodipropionitrile and rat brain aromatic transferases. *Biochem. Pharmacol.* 8:767-775.

51. Geller, A., F. Gilles, and H. Schwachman. 1977. Degeneration of the fasciculus gracilis in optic fibrosis. *Neurology* 27:185-187.

52. Ghetti, B. 1979. Induction of neurofibrillary degeneration following treatment with maytansine *in vivo*. *Brain Res.* 163:9-19.

53. Goldin, A., H. A. Noe, B. H. Larding, D. M. Shapiro, and B. Goldberger. 1948. A neurological syndrome induced by administration of some chlorinated tertiary amines. *J. Pharmacol. Exp. Ther.* 94:249-261.

54. Goldman, R. D. 1971. The role of three cytoplasmic fibers in BHK-21 cell motility: I. Microtubules and the effects of colchicine. *J. Cell Biol.* 51:752-762.

55. Goldschmidt, R. B., and O. Steward. 1980. Preferential neurotoxicity of colchicine for granule cells of the dentate gyrus of the adult rat. *Proc. Natl. Acad. Sci. USA* 77:3047-3051.

56. Goldschmidt, R. B., and O. Steward. 1982. Neurotoxic effects of colchicine: Differential susceptibility of CNS neuronal populations. *Neuroscience* 7:695-714.

57. Graham, D. G. 1980. Hexane neuropathy: A proposal for pathogenesis of a hazard of occupational exposure and inhalant abuse. *Chem. Biol. Interact.* 32:339-345.

58. Gray, E. G., and L. L. Hamlyn. 1962. Electron microscopy of experimental degeneration in the avian optic tectum. *J. Anat.* 96:309-316.

59. Green, L. S., A. Donoso, I. E. Heller-Bettinger, and F. E. Samson. 1977. Axonal transport disturbances in vincristine-induced peripheral neuropathy. *Ann. Neurol.* 1:255-262.

60. Griffin, J. W., L. C. Cork, R. M. Adams, and D. L. Price. 1982. Axonal transport in hereditary canine spinal muscular atrophy. *J. Neuropathol. Exp. Neurol.* 41:370 (abs.).

61. Griffin, J. W., R. E. Fahnstock, and D. L. Price. 1982. Fast axonal transport through

focal regions of cytoskeletal segregation produced by chemical neurotoxins. *Neurology (N.Y.) 32*:A214 (abs.).

62. Griffin, J. W., B. G. Gold, L. C. Cork, D. L. Price, and H. E. Lourdes. 1982. IDPN neuropathy in the cat: Coexistence of proximal and distal axonal swellings. *Neuropath. Appl. Neurobiol.*, in press.

63. Griffin, J. W., P. N. Hoffman, A. W. Clark, P. T. Carroll and D. L. Price. 1978. Slow axonal transport of neurofilament proteins: Impairment by β, β'-iminodipropionitrile administration. *Science 202*:633-635.

64. Griffin, J. W., and D. L. Price. 1976. Axonal transport in motor neuron pathology. *In*: Recent Research Trends. J. Andrews, R. Johnson, and M. Brazier, editors. Academic Press, New York, p. 33.

65. Griffin, J. W., and D. L. Price. 1981. Demyelination in experimental β, β'-iminodipropionitrile and hexacarbon neuropathies. *Lab. Invest. 45*:130-141.

66. Griffin, J. W., D. L. Price, and D. B. Drachman. 1977. Impaired axonal regeneration in acrylamide intoxication. *J. Neurobiol. 8*:355-370.

67. Griffin, J. W., D. L. Price, and P. S. Spencer. 1977. Fast axonal transport through giant axonal swellings in hexacarbon neuropathy. *J. Neuropathol. Exp. Neurol. 36*:603 (abs.).

68. Griffiths, I. R., and I. D. Duncan. 1979. Central nervous system in a case of giant axonal neuropathy. *Acta Neuropathol. (Berl.) 46*:169-172.

69. Hartman, H. A., and L. Murmanis. 1962. Electron microscopic alterations of spinal motor neurons produced by β-β'-iminodipropionitrile. *Fed Proc. 21*:362 (abs.).

70. Hashimoto, K., and W. N. Aldridge. 1970. Biochemical studies on acrylamide, a neurotoxic agent. *Biochem. Pharmacol. 19*:2591-2604.

71. Heath, J. W., S. Ueda, M. B. Bornstein, G. D. Daves, and C. S. Raine. 1982. Buckthorn neuropathy *in vitro*: Evidence for a primary neuronal effect. *J. Neuropathol. Exp. Neurol. 41*:204-220.

72. Herskowitz, A., N. Ishii, and H. Schaumberg. 1971. n-Hexane neuropathy. *N. Eng. J. Med. 285*:82-85.

73. Hetnarski, B., H. M. Wisniewski, K. Iqbal, J. D. Dzidzic, and A. Lajtha. 1980. Central cholinergic activity in aluminum-induced neurofibrillary generation. *Ann. Neurol. 7*:480-490.

74. Hirano, A., and H. M. Zimmerman. 1970. Some effects of vinblastine implantation in the cerebral white matter. *Lab. Invest. 23*:358-367.

75. Hoffman, P. N., and R. J. Lasek. 1975. The slow component of axonal transport: Identification of major structural polypeptides of the axon and their generality among mammalian neurons. *J. Cell Biol. 66*:351-366.

76. Holtzer, H., S. A. Fellini, N. Rubinstein, J. Chi, and K. Straks. 1976. *In*: Cell Moltility. R. Goldman, T. Pollard, and J. Rosenbaum, editors. Cold Spring Harbor, N.Y., pp. 823-840.

77. Hopkins, A. 1975. Toxic neuropathy due to industrial agents. *In*: Peripheral Neuropathy. P. J. Dyck, P. K. Thomas, and E. H. Lambert, editors. W. B. Saunders, Philadelphia, pp. 1207-1226.

78. Huber, D. T., and E. Frieden. 1970. The inhibition of ferroxidase by trivalent and other metal ions. *J. Biol. Chem. 245*:3979-3984.

79. Hynes, R. D., and A. T. Destree. 1978. 10 nm filaments in normal and transformed cells. *Cell 13*:151-163.

80. Ihara, Y., and D. J. Selkoe. Purification and partial characterization of paired helical filaments in Alzheimer's disease (submitted).

81. Ishii, N., A. Herskowitz, and H. H. Schaumberg. 1972. n-Hexane polyneuropathy: A clinical and experimental study. *J. Neuropathol. Exp. Neurol. 31*:198 (abs.).

82. Ishikawa, H., R. Bischoff and H. Holtzer. 1968. Mitosis and intermediate sized filaments in developing skeletal muscle. *J. Cell Biol. 38*:538-555.

83. Julien, J. P., W. E. Mushyushi, I. D. Duncan, and I. R. Griffith. 1981. Giant axonal

neuropathy: neurofilaments isolated from diseased dogs have a normal polypeptide composition. *Exp. Neurol. 72*:619-627.

84. Juntunen, J., M. Haltia, and I. Linnoila. 1974. Histochemically demonstrable nonspecific cholinsterase as an indicator of peripheral nerve lesion in carbon disulphide-induced polyneuropathy. *Acta Neuropathol. (Ber). 29*:361-366.

85. Kaplan, M. L., and D. Murphy. 1972. Effect of acrylamide on rotarod performance and sciatic nerve β-glucuronidase activity in rats. *Toxicol. Appl. Pharmacol. 22*:259-268.

86. Kaplan, M. L., and D. Murphy. 1972. Modifications of acrylamide neuropathy in rats by various factors. *Toxicol. Appl. Pharmacol. 22*:302-303.

87. Karlik, S. J., G. L. Eichhorn, P. N. Lewis, and D. R. Crapper. 1980. Interaction of aluminum species with deoxyribonucleic acid. *Biochemistry 19*:5991-5998.

88. Kidd, M. 1963. Paired helical filaments in electron microscopy in Alzheimer's disease. *Nature 197*:192-193.

89. Kim, S. V., M. Thomonaga, and B. Ghetti. 1980. Neurofibrillary degeneration in cultured adult mouse neurons induced by maytansine. *Acta Neuropathol. (Berl.) 52*:161-164.

90. King, G., U. DeBoni, and D. R. Crapper. 1975. Effect of aluminum upon conditioned avoidance response acquisition in the absence of neurofibrillary degeneration. *Pharmacol. Biochem. Behav. 3*:1003-1009.

91. Kitiyakara, A., M. C. Shively, and H. A. Hartmann. 1977. Synthesis and fate of RNA in neurons from mice given β, β'-iminodipropionitrile. *Exp. Neurol. 54*:403-413.

92. Klatzo, I., H. Wisniewski, and E. Streicher. 1965. Experimental production of neurofibrillary degeneration: 1. Light microscopic observation. *J. Neuropath. Exp. Neurol. 24*: 187-199.

93. Kopeloff, L. M., S. E. Barrera, and N. Kopeloff. 1942. Recurrent convulsive seizures in animals produced by immunogenic and chemical means. *Am. J. Psychiat. 98*:881-902.

94. Korobkin, B., A. K. Asbury, A. J. Sumner, and S. L. Nielsen. 1975. Glue-sniffing neuropathy. *Arch. Neurol. 32*:158-162.

95. Kosik, K. S., C. Rasool, W. G. Bradley, and D. J. Selkoe. 1982. Increased rate of accumulation of ChAT proximal to a sciatic nerve ligature in Al^{3+}-treated rabbits. *J. Neuropathol. Exp. Neurol. 41*:361 (abs.).

96. Krasavage, W. J., J. L. O'Donoghue, G. D. DiVincenzo, and C. J. Terhaar. 1980. The relative neurotoxicity of methyl-n-butyl ketone, n-hexane, and their metabolites. *Toxicol. Appl. Pharmacol. 52*:433-441.

97. Kreutzberg, G. 1969. Neuronal dynamics and axonal flow: IV. Blockage of intra-axonal enzyme transport by colchicine. *Proc. Natl. Acad. Sci. USA 62*:722-728.

98. Kuperman, A. S. 1958. Effects of acrylamide on the central nervous system of the cat. *J. Pharmacol. Exp. Ther. 123*:180-192.

99. Lampert, P., J. M. Blumberg, and A. Pentschew. 1964. An electron microscopic study of dystophic axons in the gracile and cuneate nuclei of vitamin E-deficient rats. *J. Neuropathol. Exp. Neurol. 23*:60-77.

100. Langlais, P. J., P. C. Huang, and S. Gabay. 1975. Regional neurochemical studies on the effect of β, β'-iminodipropionitrile (IDPN) in the rat. *J. Neurosci. Res. 1*:419-535.

101. Lee, J. C-Y. 1963. Electron microscopy of Wallerian degeneration. *J. Comp. Neurol. 120*:65-71.

102. Linnoila, I., M. Haltia, A-M. Seppäläinen, and J. Palo. 1975. Experimental carbon disulphide poisoning: Morphological and neurophysiological studies. *In*: Proceedings, VIIth International Congress of Neuropathology, Budapest, 1974. Vol. 2. St. Környey, St. Tariska, and G. Gosztonyi, editors. Excerpta Medica, Amsterdam, pp. 383-386.

103. Liss, L., K. Ebner, D. Couri, and N. Cho. 1975. Alzheimer disease: A possible animal model. IV Panamerican Congress of Neurology, Mexico.

104. Liwincz, B. H., K. Kristensson, H. M. Wisniewski, M. L. Shelanski, and R. D. Terry. 1974. Observations on axoplasmic transport in rabbits with aluminum-induced neurofibrillary tangles. *Brain Res. 80*:413-420.

190   Kosik and Selkoe

105. McCollister, D. D., F. Oyen, and V. K. Rowe. 1964. Toxicology of acrylamide. *Toxicol. Appl. Pharmacol. 6*:172-181.

106. Madelbaum-Shavit, F., M. R. Wolpert-DeFilippes, and D. G. Johns. 1976. Binding of maytansine to rat brain tubulin. *Biochem. Biophys. Res. Commun. 12*:47-54.

107. Mäkinen, A., H. Savolainen, E. Lehtonen, and H. Vanio. 1977. Reduced sulfhydryl groups of rat neurons, glial cells and neurofilaments. *Res. Commun. Chem. Path. Pharm. 16*: 577.

108. Martin, J. J., and L. Martin. 1972. Infantile neuroaxonal dystrophy: Ultrastructural study of the peripheral nerves and motor end plates. *Eur. Neurol. 8*:239-250.

109. Mendell, J. R., K. Saida, M. F. Ganansia, D. B. Jackson, H. S. Weiss, R. W. Gardier, C. Christman, N. Alen, D. Couri, J. O'Neil, B. Marks, and L. Hetland. 1974. Toxic polyneurothpathy produced by methyl n-butyl ketone. *Science 185*:787-789.

110. Mendell, J. R., S. Zarife, K. Saida, H. S. Weiss, R. Savage, and D. Couri. 1977. Alterations of fast axoplasmic transport in experimental methyl n-butyl ketone neuropathy. *Brain Res. 133*:107-118.

111. Mennin, S., and D. W. Thomas. 1970. Comparative effects of an osteolathyrogen and a neurolathyrogen on brain and connective tissue. *Proc. Soc. Exp. Biol. Med. 134*:489-491.

112. Miller, C. A., and E. M. Levine. 1974. Effects of aluminum salts on cultured neuroblastoma cells. *J. Neurochem. 22*:751-758.

113. Moddell, G., J. M. Bilbao, D. Payne, and P. Ashby. 1978. Disulfiram neuropathy. *Arch. Neurol. 35*:658-660.

114. Mori, H., and M. Kurokawa. 1979. Purification of neurofilaments and their interaction with vinblastine sulphate. *Cell Struct. Funct. 4*:163-167.

115. Nelson, J. S., C. D. Fitch, V. W. Fischer, G. O. Brown, and A. C. Chou. 1981. Progressive neuropathologic lesions in vitamin E-deficient rhesus monkeys. *J. Neuropathol. Exp. Neurol. 40*:166-186.

116. Niklowitz, W. J. 1975. Neurofibrillary changes after acute experimental lead poisoning. *Neurology 25*:927-934.

117. O'Donoghue, J. L., and W. J. Krasavage. 1979. Hexacarbon neuropathy: A γ-diketone neuropathy? *J. Neuropathol. Exp. Neurol. 38*:333(abs.).

118. Ohmi, S. 1961. Electron microscopic study on Wallerian degeneration of the peripheral nerves. *Z. Zellforsch. 54*:39-67.

119. O'Shea, K. S., and M. H. Kaufman. 1980. Copper-induced microtubule degeneration and filamentous inclusions in the neuroepithelium of the mouse embryo. *Acta Neuropathol. (Berl.) 49*:237-240.

120. Papasozemenos, S. Ch., L. Autilio-Gambetti, and P. Gambetti. 1981. Reorganization of axoplasmic organelles following β, β'-iminodipropionitrile administration. *J. Cell Biol. 91*:866-871.

121. Papasozemenos, S. Ch., L. Autilio-Gambetti, and P. Gambetti. 1982. Redistribution of proteins of fast axonal transport in β, β'-iminodipropionitrile (IDPN) intoxication: A quantitative electron microscope autoradiographic (EMA) study. *J. Neuropathol. Exp. Neurol. 41*:371 (abs.).

122. Parhad, I. M., A. W. Clark, J. S. Folus, J. W. Griffin, and D. L. Price. 1982. Neurofilamentous axonal swellings in doxorubicin-induced retinal neuropathy. *J. Neuropathol. Exp. Neurol. 41*:358 (abs.).

123. Parhad, I. M., J. W. Griffin, D. L. Price, A. W. Clark, L. C. Cork, and N. R. Miller. 1982. Intoxication with β, β'-iminodipropionitrile (IDPN): A model of optic disc swelling. *Lab. Invest.*, in press.

124. Paulson, J. C., and W. O. McClure. 1974. Microtubules and axoplasmic transport. *Brain Res. 73*:333-337.

125. Paulson, J. C., and W. O. McClure. 1975. Inhibition of axoplasmic transport by colchicine, podophyllotoxin and vinblastine: An effect on microtubules. *Ann. N.Y. Acad. Sci. 253*:517-527.

126. Peiffer, I., W. Schlote, A. Bischoff, E. Boltshauser, and G. Muller. 1977. Generalized giant axonal neuropathy: A filament-forming disease of neuronal, endothelial, glial and Schwann cells in a patient without kinky hair. *Acta Neuropathol.* 40:213-218.

127. Pena, S. D. J. 1982. Giant axonal neuropathy: An inborn error of organization of intermediate filaments. *Muscle & Nerve* 5:166-172.

128. Pentschew, A., and K. Schwartz. 1962. Systemic axonal dystrophy in vitamin E deficient adult rats. *Acta Neuropathol. (Berl.)* 1:313-334.

129. Perl, D. P., and A. R. Brody. 1980. Alzheimer disease: X-ray spectrometric evidence of aluminum accumulation in neurofibrillary tangle-bearing neurons. *Science 208*: 297-299.

130. Peterson, E. R., and M. B. Bornstein. 1968. The neurotoxic effects of colchicine on tissue cultures of cord-ganglia. *J. Neuropathol. Exp. Neurol.* 27:121 (abs.).

131. Peterson, E. R., and M. R. Murray. 1966. Serial observations in tissue cultures of neurotoxic effects of colchicine. *Anat. Rec.* 154:401 (abs.).

132. Petit, T. L., G. B. Biederman, and P. S. McMullin. 1980. Neurofibrillary degeneration, dendritic dying back, and learning-memory deficits after aluminum administration: Implications for brain aging. *Exp. Neurol.* 67:152-162.

133. Potter, H. D., and G. S. Hafner. 1974. Sequence of changes in neurofibrils (neurofilaments) induced in synaptic regions of bullfrogs by environmental temperature changes. *J. Comp. Neurol.* 155:409-424.

134. Potter, H. D., G. S. Hafner, and I. R. Schwartz. 1975. Neurofilament and glycogen changes during cold acclimation on the trochlear nucleus of lizards (*Sceloporus undulatus*). *J. Neurocytol.* 4:491-503.

135. Prineas, J. 1969. The pathogenesis of dying back polyneuropathies: Part 1. An ultrastructural study of experimental tri-orthocresyl phosphate intoxication in the cat. *J. Neuropathol. Exp. Neurol.* 28:571-597.

136. Prineas, J. 1969. The pathogenesis of dying back polyneuropathies: Part II. An ultrastructural study of experimental acrylamide intoxication in the cat. *J. Neuropathol. Exp. Neurol.* 28:598-621.

137. Raine, C. S., G. Ghetti, and M. L. Shelanski. 1971. On the association between tubules and mitochondria in axons. *Brain Res.* 34:389-393.

138. Rasool, C. G., and W. G. Bradley. 1978. Axonal transport in acrylamide neuropathy. *In*: Peripheral Neuropathies. N. Canal and G. Pozza, editors. Elsevier/North Holland Biomedical Press, Amsterdam, pp. 159-165.

139. Rasool, C. G., and W. G. Bradley. 1978. Studies on axoplasmic transport of individual proteins: 1-acetylcholinesterase (AChE) in acrylamide neuropathy. *J. Neurochem.* 31:419-425.

140. Remillard, S., L. I. Rebhun, G. A. Howie, and S. M. Kupchan. 1975. Antimitotic activity of the potent tumor inhibitor maytansine. *Science 189*:1002-1005.

141. Robbins, E., and N. K. Gonatas. 1964. Histochemical and ultrastructural studies on HeLa cell cultures exposed to spindle inhibitors with special reference to the interphase cell. *J. Histochem. Cytochem.* 12:704-711.

142. Roots, B. I., and R. L. Bordar. 1977. Neurofilamentous changes in goldfish (*Carrasius aurautus* L.) brain in relation to environmental temperature. *J. Neuropathol. Exp. Neurol.* 36:453-464.

143. Runge, M. S., T. M. Lane, D. A. Yphantis, M. R. Lifsics, A. Saito, M. Altin, K. Reinke, and R. C. Williams, Jr. 1981. ATP-induced formation of an associated complex between microtubules and neurofilaments. *Proc. Natl. Acad. Sci. USA* 78:1431-1435.

144. Sabri, M. I., C. L. Moore, and P. S. Spencer. 1979. Studies on the biochemical basis of distal axonopathies: I. Inhibition of glycolysis by neurotoxic hexacarbon compounds. *J. Neurochem.* 32:683-689.

145. Saida, K., J. R. Mendell, and H. S. Weiss. 1976. Peripheral nerve changes induced by methyl n-butyl ketone and potentiation by methyl ethyl ketone. *J. Neuropathol. Exp. Neurol.* 35:207-225.

146. Samson, F. E. 1971. Mechanism of axoplasmic transport. *J. Neurobiol.* 2:347-360.

147. Sanderson, C. J., D. R. Crapper, and U. DeBonni. 1979. Altered response to ecdysterone by chromatin-bound aluminum in a polytene chromosome of simulium vittatum. *J. Cell Biol. 83*:152 (abs.).

148. Sato, Y., S. V. Kim, and B. Ghetti. 1982. Neurofibrillary tangle formation in cultured neurons. *J. Neuropathol. Exp. Neurol. 41*:341 (abs.).

149. Savolainen, H. 1977. Some aspects of the mechanisms by which industrial solvents produce neurotoxic effects. *Chem. Biol. Inter. 18*:1-10.

150. Savolianen, H., E. Lehtonen, and H. Vanio. 1977. CS_2 binding to rat spinal neurofilaments. *Acta Neuropathol. 37*:219-223.

151. Scherp, H. W., and C. Church, 1937. Neurotoxic action of aluminum salts. *Proc. Soc. Exp. Biol. (N.Y.) 36*:851-853.

152. Schlaepfer, W. W. 1971. Cadmium injury in rat spinal ganglia *in vivo. J. Neuropathol. Exp. Neurol. 30*:141-157.

153. Schochet, S. S., Jr., P. W. Lampert, and K. M. Earle. 1968. Neuronal changes induced by intrathecal vincristine sulfate. *J. Neuropathol. Exp. Neurol. 27*:645-658.

154. Schor, N. A., K. Archer, and H. A. Hartmann. 1971. Enzyme activities in the anterior horn of spinal cord after β, β'-iminodipropionitrile. *Exp. Neurol. 33*:351-359.

155. Seibert, B., and H. Wells. 1929. The effect of aluminum on mammalian blood and tissues. *Arch. Pathol. 8*:230-262.

156. Seil, F. J., and J. Lampert. 1968. Neurofibrillary tangles induced by vincristine and vinblastine sulfate in central and peripheral neurons *in vitro. Exp. Neurol. 21*:219-230.

157. Selkoe, D. J., Y. Ihara, and F. J. Salazar. 1982. Alzheimer's disease: Insolubility of partially purified paired helical filaments in sodium dodecyl sulfate and urea. *Science 215*: 1243-1245.

158. Selkoe, D. J., R. K. Liem, S. H. Yen, and M. L. Shelanski. 1979. Biochemical and immunological characterization of neurofilaments in experimental neurofibrillary degeneration induced by aluminum. *Brain Res. 163*:235-252.

159. Selkoe, D. J., L. Luckenbill-Edds, and M. L. Shelanski. 1978. Effects of neurotoxic industrial solvents on cultured neurobastoma cells: Methyl n-butyl ketone, n-hexane and derivatives. *J. Neuropathol. Exp. Neurol. 37*:768-789.

160. Selye, H. 1957. Lathyrism. *Revue Canad. Biol. 16*:1.

161. Sharer, L. R., and H. E. Loundes. 1982. Acrylamide-induced ascending generation of ligated peripheral nerve. *J. Neuropathol. Exp. Neurol. 41*:348 (abs.).

162. Shelanski, M. L., J-F. Letterier, and R. K. H. Liem. 1979. Evidence for interactions between neurofilaments and microtubules. *Neurosci. Res. Prog. Bull. 19*:32-43.

163. Shelanski, M. L., and H. Wisneiwski. 1969. Neurofibrillary degeneration induced by vincristine therapy. *Arch. Neurol. 29*:199-205.

164. Shimono, M., K. Izumi, and Y. Kurowa. 1978. β, β'-iminodipropionitrile induced centrifugal segmented demyelination and onion bulb formation. *J. Neuropathol. Exp. Neurol. 37*:375-386.

165. Shirabe, T., T. Tsuda, and A. Terao. 1974. Toxic polyneuropathy due to glue-sniffing: Report of two cases with light and electron microscopic study of the peripheral nerves and muscles. *J. Nuerol. Sci. 21*:101-113.

166. Siem. 1886. *In*: Dölken; 1898. Über die Wirkung des Aluminum, mit besonderer Berücksichtigung der durch das Aluminum versursacthen Läsionen im Zentralnervensystem. *Naunyn-Schmiederbergs Arch. Exp. Path. Pharmak. 40*:99.

167. Slagel, D. E., and H. A. Hartmann. 1965. The distribution of neuroaxonal lesions in mice injected with iminodipropionitrile with special reference to the vestibular system. *J. Neuropathol. Exp. Neurol. 24*:599-620.

168. Soifer, D., K. Iqbal, H. Czosnek, J. DeMartini, J. A. Sturman, and H. M. Wisneiwski. 1981. The loss of neuron-specific proteins during the course of Wallerian degeneration of optic and sciatic nerve. *J. Neurosci 1*:461-470.

169. Souyri, F., M. Chretien, and B. Droz. 1981. "Acrylamide-induced" neuropathy and impairment of axonal transport of proteins: I. Multifocal retention of fast transported proteins at the periphery of axons as revealed by light microscope radioautography. *Brain Res.* 205:1-13.

170. Spencer, P. S., D. Couri and H. H. Schaumberg. 1980. n-Hexane and methyl n-butyl ketone. *In*: Experimental and Clinical Neurotoxicology. P. S. Spencer and H. H. Schaumberg, editors. Williams and Wilkins, Baltimore, pp. 456-475.

171. Spencer, P. S., M. I. Sabri, H. H. Schaumberg, and C. L. Moore. 1979. Does a defect of energy metabolism in the nerve fiber underlie axonal degeneration in polyneuropathies. *Ann. Neurol.* 5:501-507.

172. Spencer, P. S., and H. H. Schaumberg. 1975. Experimental neuropathy produced by 2,5-hexane-dione: A major metabolite of the neurotoxic industrial solvent methyl n-butyl ketone. *J. Neurol. Neurosurg. Psych.* 38:771-775.

173. Spencer, P. S., and H. H. Schaumberg. 1976. Towards a molecular basis for experimental giant axonal degeneration. *J. Neuropathol. Exp. Neurol.* 35:349 (abs.).

174. Spencer, P. S., and H. H. Schaumberg. 1977. Neurotoxic properties of certain aliphatic hexacarbons. *Proc. Roy. Soc. Med.* 70:37-39.

175. Spencer, P. S., and H. H. Schaumberg. 1977. Ultrastructural studies of the dying-back process: III. The evolution of experimental peripheral giant axonal degeneration. *J. Neuropathol. Exp. Neurol.* 36:276-299.

176. Spencer, P. S., and H. H. Schaumberg. 1977. Ultrastructural studies of the dying-back process: IV. Differential vulnerability of the PNS and CNS fibers in experimental central-peripheral distal axonopathies. *J. Neuropathol. Exp. Neurol.* 36:300-320.

177. Spencer, P. S., H. H. Schaumberg, and E. R. Peterson. 1975. Neurofilamentous axonal degeneration *in vivo* and *in vitro* produced by 2,5-hexandione. *Neurosci. Abs.* 1:703.

178. Spencer, P. S., H. H. Schaumberg, R. L. Raleigh, and C. J. Terhaar. 1975. Nervous system degeneration produced by the industrial solvent methyl n-butyl ketone. *Arch. Neurol.* 32:219-222.

179. Spencer, P. S., H. H. Schaumberg, M. I. Sabri, and B. Verones. 1980. CRC Critical Reviews in Toxicology. Vol. 7, Issue 4. CRC Press, Boca Raton, Fla., pp. 279-357.

180. Spencer, P. S., and P. K. Thomas. 1974. Ultrastructural studies of the dying-back process: II. The sequestration and removal by Schwann cells and oligodendrocytes of organelles from normal and diseased axons. *J. Neurocytol.* 3:763-783.

181. Starger, J. M., and R. D. Goldman. 1977. Isolation and preliminary characterization of 10-nm filaments from baby hamster kidney (BHK-21) cells. *Proc. Natl. Acad. Sci. USA* 74:2422-2426.

182. Sterman, A. B. 1982. Distinct ultrastructural reorganization of the nerve cell body in acrylamide neuropathy. *J. Neuropathol. Exp. Neurol.* 41:361 (abs.).

183. Sumner, A., D. Pleasure and K. Ciesielka. 1976. Slowing of fast axoplasmic transport in acrylamide neuropathy. *J. Neuropathol. Exp. Neurol.* 35:319 (abs.).

184. Sung, J. H. 1964. Neuroaxonal dystrophy in mucoviscidosis. *J. Neuropathol. Exp. Neurol.* 23:567-583.

185. Sung, J. H., and E. M. Stadlan. 1966. Neuroaxonal dystrophy in congenital biliary atresia. *J. Neuropathol. Exp. Neurol.* 25:341-361.

186. Szendyskowski, S., J. Stetkiewicz, T. Wronska-Nofer, and M. Karasek. 1974. Pathomorphology of the experimental lesion of the peripheral nervous system in white rats chronically exposed to carbon disulphide. *In*: Structure and Function of Normal and Diseased Muscle and Peripheral Nerve. I. Hausmanowa-Petrusewicz and H. Jedrzejowska, editors. Polish Medical Publishers, Warsaw.

187. Taylor, E. W. 1965. The mechanism of colchicine inhibition of mitosis: I. Kinetics of inhibition and the binding of ^{3}H-colchicine. *J. Cell Biol.* 25:145-160.

188. Tello, F. 1904. Las neurofibrillas en los vertebrados inferiores. *Trab. Lab. Invest. Biol.* 3:113-151.

189. Terry, R. D., and C. Pena. 1965. Experimental production of neurofibrillary degeneration: 2. Electron microscopy, phosphatase histochemistry and electron probe analysis. *J. Neuropathol. Exp. Neurol. 24*:200-210.

190. Tischner, K. H., and J. M. Schroder. 1972. The effects of cadmium chloride on organotype cultures of rat sensory ganglia: A light and electron microscopic study. *J. Neurol. Sci. 16*:383-399.

191. Trapp, G. A. 1980. Studies of aluminum interaction with enzymes and proteins: The inhibition of hexokinase. *In:* Aluminum Neurotoxicity. L. Liss, editor. Pathotox Publishers, Park Forest South, Ill., pp. 89-100.

192. Troncoso, J. C., D. L. Price, J. W. Griffin, and I. M. Parhad. 1982. Neurofibrillary axonal pathology in aluminum intoxication. *Neurology 32*:227.

193. Troncoso, J. C., E. F. Stanley, J. W. Griffin, D. L. Price, and L. C. Cork. 1982. Canine neuroaxonal dystrophy. *J. Neuropathol. Exp. Neurol. 41*:363 (abs.).

194. Ule, G. 1961. Experimenteller Neurolathyrismus. *Verh. Dtsch. Ges. Path. 45*:333-338.

195. Ule, G. 1962. Zur Ultrastruktur der Ghost cells beim experimentellen Neurolathyrismus der Ratte. *Z. f. Zellforsch. 56*:130-142.

196. Vandervelde, M., C. E. Greene, and E. J. Hoff. 1976. Lower motor neuron disease with accumulation of neurofilaments in a cat. *Vet. Pathol. 13*:428-435.

197. Veronesi, B., E. R. Peterson, G. DiVincenzo, and P. S. Spencer. 1978. A tissue culture model of distal (dying-back) axonopathy: Its use in determining primary neurotoxic hexacarbon compounds. *J. Neuropathol. Exp. Neurol. 37*:703 (abs.).

198. Virtanen, I., V. P. Lehto, and E. Lehtonen. 1980. Organization of intermediate filaments in cultured fibroblasts upon disruption of microtubules by cold treatment. *Eur. J. Cell Biol. 23*:80-84.

199. Volk, B. 1980. Paired helical filaments in rat spinal ganglia following chronic alcohol administration: An electron microscopic investigation. *Neuropath. Appl. Neurobiol. 6*: 143-153.

200. Webster, H. de F. 1962. Transient focal accumulation of axonal mitochondria during the early stages of Wallerian degeneration. *J. Cell Biol. 12*:361-383.

201. Wenk, G. L., and K. K. Stemmer. 1982. Activity of the enzymes dopamine-beta-hydroxylase and phenylethanolamine-N-methyltransferase in discrete brain regions of the copper-zinc deficient rat following aluminum ingestion. *Neurotox. 3*:93-99.

202. Wisniewski, H. M., M. E. Bruce, and H. Fraser. 1975. Infectious etiology of neuritic (senile) plaques in mice. *Science 190*:1108-1110.

203. Wisniewski, H., O. Narkiewicz, and K. Wisniewska. 1967. Topography and dynamics of neurofibrillary degeneration in aluminum encephalopathy. *Acta Neuropathol (Berl.) 9*: 127-133.

204. Wisniewski, H., M. Shelanski, and R. D. Terry. 1968. Effects of mitotic spindle inhibitors on neurotubules and neurofilaments in anterior horn cells. *J. Cell Biol. 38*:224-229.

205. Wisniewski, H. M., and D. Soifer. 1979. Neufoibrillary pathology: Current status and research perspectives. *Mech. Aging Devel. 9*:119-142.

206. Wisniewski, H. M, J. A. Sturman, and J. W. Shek. 1980. Aluminum chloride induced neurofibrillary changes in the developing rabbit: A chronic animal model. *Ann. Neurol. 8*:479-490.

207. Wisniewski, H., and R. D. Terry. 1967. Experimental colchicine encephalopathy: I. Induction of neurofibrillary degeneration. *Lab. Invest. 17*:577-587.

208. Wisniewski, H. M., G. Y. Wen, and A. A. Lidsky. 1978. Aluminum-induced neurofibrillary changes in rabbit retina: ERG and morphological studies. *In:* Proceedings, VIII International Congress of Neuropathology. Washington, D. C., p. 708.

209. Wood, P. L., and R. J. Boegman. 1975. Increased rate of rapid axonal transport in vitamin E deficient rats. *Brain Res. 84*:325-328.

210. Yamamura, Y. 1969. n-Hexane polyneuropathy. *Folia Psychiatr. Neurol. Jap. 23*: 45-47.

211. Yates, C. M., A. Gordon, and H. Wilson. 1976. Neurofibrillary degeneration induced in the rabbit by aluminum chloride: Aluminum neurofibrillary tangles. *Neuropath. Appl. Neurobiol. 2*:131-144.

212. Yates, C. M., J. Simpson, D. Russell, and A. Gordon. 1980. Cholinergic enzymes in neurofibrillary degeneration produced by aluminum. *Brain Res. 197*:269-274.

213. Yen, S-H., D. Dahl, M. Schachner, and M. L. Shelanski. 1976. Biochemistry of the filaments of brain. *Proc. Natl. Acad. Sci. USA 73*:529-533.

214. Yokoyama, K., S. Tsukita, H. Ishikawa, and M. Kurokawa. 1980. Early changes in the neuronal cytoskeleton caused by β, β'-iminodipropionitrile: Selective impairment of neurofilament polypeptides. *Biomed. Res.* 537-547.

Chapter 7

Neurofibrillary Tangles and Paired Helical Filaments in Alzheimer's Disease

Henryk M. Wisniewski, George S. Merz, Patricia A. Merz, Guang Y. Wen, and Khalid Iqbal

Alzheimer's Disease

Alzheimer's disease (AD) is a degenerative disease of the human central nervous system (CNS) that is characterized by a progressive loss of mental functioning that culminates in a profound dementia. Although the term was originally used to describe a specific form of presenile dementia, it is now clear that the disease occurs much more frequently in the senium. Arbitrarily, the disease is labeled presenile dementia or Alzheimer's disease when it occurs prior to age 65 and senile dementia of the Alzheimer type (SDAT) after age 65.

The impact of SDAT on the elderly has only recently become appreciated. Of the 25 million in the United States population currently over 65 years old about 10%, or 2.5 million, exhibit some cognitive deficits. At least half of these individuals are afflicted with dementia severe enough to require nursing home care (29). Approximately three-quarters of these severe cases are SDAT, which makes it the largest single cause of dementia in the aged. Although vascular disease has long been held as the major cause of dementia (i.e., multiinfarct dementia), in actuality it accounts for less than a quarter of these severe cases. Furthermore, demographic projections by the U.S. Census Bureau indicate that the magnitude of the problem will increase steadily since the number of individuals over age 65 is expected to increase from 25 million in 1980 to about 55 million by 2030. The vista is even bleaker when one considers that, at this moment, neither the etiology or pathogenesis of the disease is known nor is there any effective means of prevention, therapy, or cure.

Pathological Alterations in Alzheimer's Disease

Histopathological examination of the postmortem Alzheimer brain reveals a broad array of morphological changes, including neurofibrillary tangles, neuritic plaques, accumulations of lipofuscin, granulovacuolar degeneration, Lewy bodies, Hirano bodies, Wallerian degeneration, and loss of dendritic spines (51). Most of these changes are also seen in normal aged human brains. However, the most prominent lesions, the neurofibrillary tangles and neuritic/amyloid plaques, are found in much higher numbers in AD/SDAT, which suggests that they play an important but as yet undefined role in the pathogenesis of Alzheimer-type dementia. Their distribution in the aged brain is largely confined to the hippocampus; in Alzheimer's disease, they are found throughout the cortex as well. Their role in the pathogenesis of dementia was much strengthened by the prospective studies of Tomlinson and others (42), in which the results of psychometric testing of mental functioning were compared with the number of lesions in an individual's brain at autopsy. It was found that the number of neurofibrillary tangles and neuritic plaques is highly correlated with the degree of dementia. It should be emphasized, however, that it is not known whether the lesions per se are the cause of the dementia or whether the lesions in conjunction with other processes together cause dementia.

In addition to the morphological change, it is now well established that there are substantial changes in neurotransmitter biochemistry. The most prominent is a large reduction in the levels of choline acetyltransferase and acetylcholinesterase in the Alzheimer brain (10, 34). Other transmitter deficits are present, but none is as prominent as those of the cholinergic system (3,10,34).

A possible relationship between the morphological and neurochemical derangements is suggested by the finding that the nucleus basalis, which sends cholinergic projections into the cortex, shows a loss of a high percentage of its cell bodies (44,45). Thus, it may be that the loss of cells in the nucleus basalis may be enough to account for the cholinergic deficits in the cortex and hippocampus of the AD/SDAT brain. The relationship, if there is one, between the cell loss in the nucleus basalis and neurofibrillary tangles is not known. It has been hypothesized, however, that the dystrophic and degenerating neurites found in the other major AD/SDAT lesion, the neuritic plaques, consist, at least in part, of cholinergic presynaptic axons arising from the basal forebrain. In their study, Struble and others (41) examined the relationship between the amount of amyloid, the degree of neuritic involvement, and the amount of acetylcholinesterase activity in plaques of aged rhesus monkeys. They found an inverse relation between the level of enzyme activity and the amount of plaque amyloid.

Because a comprehensive discussion of both neuritic and amyloid plaques is beyond the scope of this volume and because little if anything is known about how the genesis of amyloid might relate to normal cellular fibrils, the balance of this review will focus almost exclusively on paired helical filaments (PHFs) and the neurofibrillary tangle. For those interested, recent reviews of the neurotransmitter alterations and the neuritic and amyloid plaques are available (3,49).

Cellular Topography of Neurofibrillary Tangles and Paired Helical Filaments

In Bodian-stained sections of affected brain, the neurofibrillary tangle appears as an intraneuronal mass of silver-impregnated material (Figure 7-1). When stained with congo red, the Alzheimer tangles exhibit a characteristic green birefringence when viewed with polarized light.

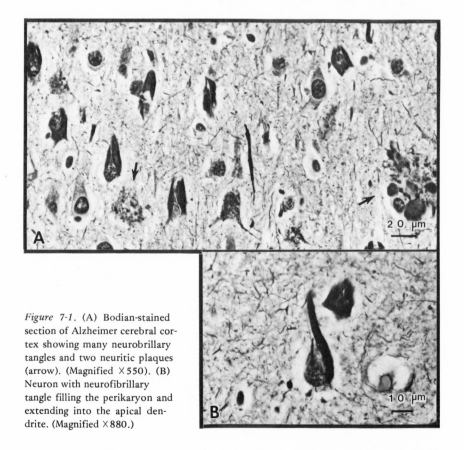

Figure 7-1. (A) Bodian-stained section of Alzheimer cerebral cortex showing many neurobrillary tangles and two neuritic plaques (arrow). (Magnified ×550). (B) Neuron with neurofibrillary tangle filling the perikaryon and extending into the apical dendrite. (Magnified ×880.)

This birefringence of Alzheimer tangles is very characteristic in that the tangles experimentally induced with aluminum do not exhibit this feature in routine paraffin sections. Electron-microscopic examination

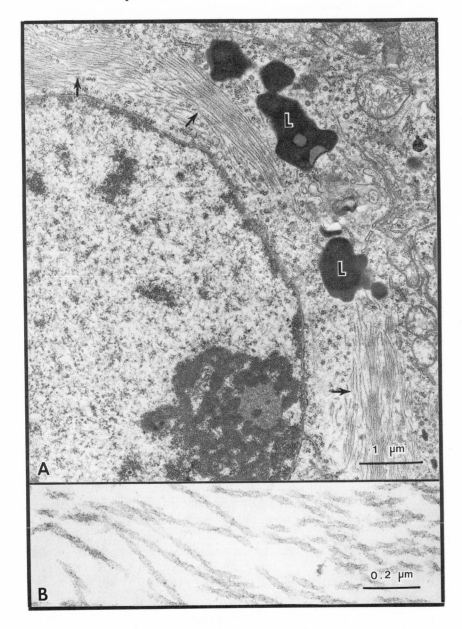

Figure 7-2. (A) Section of a neuron from Alzheimer brain with neurofibrillary tangle of PHFs (arrows) and lipofuscin granules (L). Note the relative intactness of the cellular organelles. (Magnified × 27,300). (B) High magnifications of PHFs. (Magnified × 119,500.)

reveals that the Alzheimer-type tangles are composed of PHFs with some admixture of straight 10 to 15 nm profiles (Figures 7-2 and 7-4).

The tangles of PHFs usually fill the perikaryon and appear to extend into some axons and dendrites for undetermined distances (Figures 7-1B and 7-5). At this point, we would like to stress that ultrastructural studies revealed that PHFs also occur as small clusters or single profiles both in the perikaryon and the neuritic processes (Figure 7-4). Other studies have shown that PHFs are also found in the distal portion of neurites, particularly those that are dystrophic (Figure 7-3). Although the prominent perkaryl localization of PHFs suggests that they are formed there, it is also possible that they may be formed locally in the neurites since in some instances the PHFs can also be found in neurites in areas where the surrounding perikarya are devoid of PHFs.

The effect of PHFs on the cell body and its processes is unknown. On the one hand, they are found in close association with dystrophic and/or degenerating neurites (Figure 7-3); yet, on the other, their accumulation in the perikaryon does not seem to affect other cytoplasmic organelles (Figure 7-2).

Occurrence of Paired Helical Filaments Outside of Alzheimer-Type Diseases

PHF-bearing tangles are found in a wide variety of human CNS diseases (52); this suggests that the pathological formation of PHFs reflects the limited means of the CNS to respond to a diversity of insults. With the exception of few diseases (e.g., postencephalitic Parkinsonism), however, the presence of many neurons with tangles is not a prominent feature of brain pathology. In some cases, such as aged monkeys (46), the presence of tangles is only a rare, isolated observation. On the other hand, PHFs were prominent lesions in spinal ganglia of rats after chronic ethanol ingestion (43). Finally, PHFs have also been reported to occur naturally in the large axons of whip spiders (12). All of these nonhuman forms of PHFs differ from PHFs of the Alzheimer type, principally in the distance between crossovers. To date, no one has managed to induce Alzheimer-type PHFs experimentally. The reports that PHFs are formed in explants of human and rabbit cerebral cortex after exposure to extracts of Alzheimer brain (4,11) are intriguing. However, these PHFs are ultrastructurally different from the Alzheimer-type PHFs (49).

Ultrastructure of Paired Helical Filaments

The term "paired helical filaments" (PHFs) was coined by Kidd (30). This followed from his study of negative-stained PHFs that showed

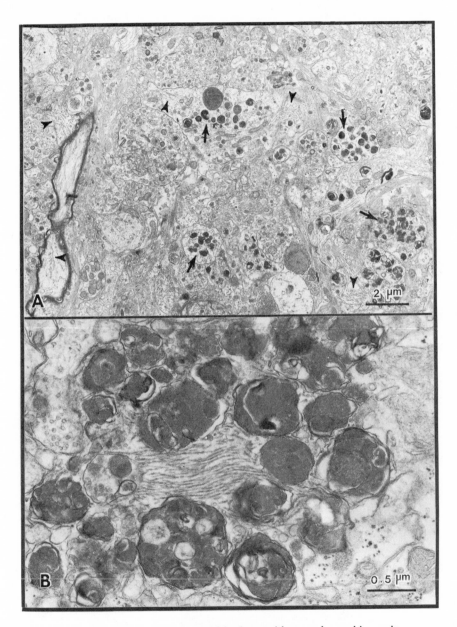

Figure 7-3. (A) Section of a primitive neuritic plaque with many dystrophic neurites with degenerating mitochondria and electron-dense bodies (arrows). Note small clusters of PHFs (points). (Magnified ×8,900.) (B) Enlargement of the dystrophic neurites with electron-dense bodies surrounding a cluster of PHFs. (Magnified ×43,000.)

two filaments that were apparently helically wound around each other. During the ensuing years, the ultrastructural characterization

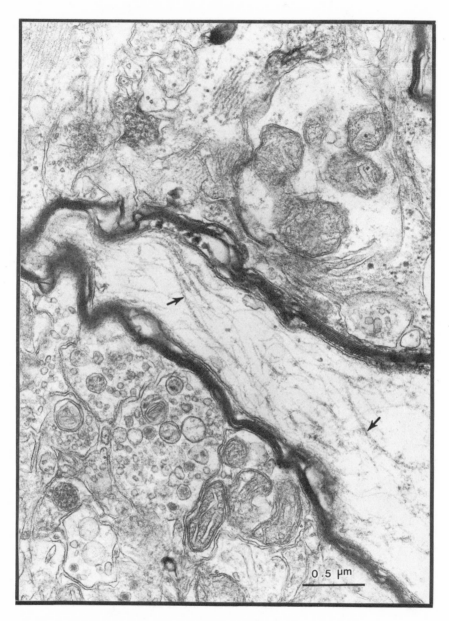

Figure 7-4. Myelinated axon with a few PHFs (arrows) mixed with neurofilaments and microtubules. (Magnified ×53,700.)

of PHFs has been pursued for the most part with thin-section electron microscopy. Using this approach, PHFs have been found to be a pair of helically wound filaments (48). The thin-section morphology

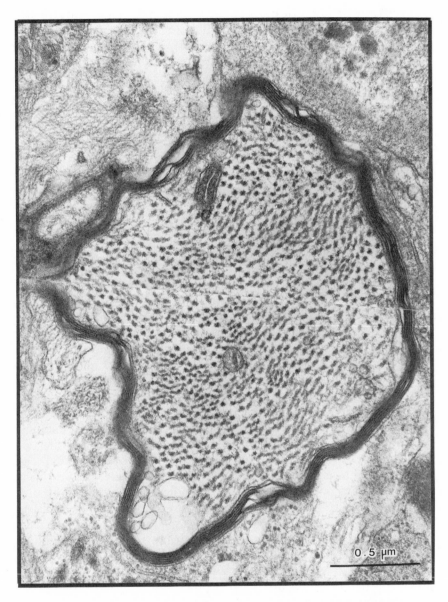

Figure 7-5. Myelinated axon filled with tangentially cut PHFs. (Magnified X 77,000.) Note lack of neurofilaments and microtubules as well as the normal appearance of the myelin.

indicated that each of these filaments might be composed of a filament 10 nm in diameter, helically wound around another filament, with an apparent crossover every 80 nm. Purification of the neurofilaments and subsequent identification of their component polypeptides (35, 36,37) has spurred a number of immunological studies of their possible relatedness to PHFs (see below). This approach was taken in part because of the difficulty encountered in purifying and solubilizing the constituent polypeptides of the PHFs. Other investigators have applied degradative procedures to neurofilaments as a means of determining their infrastructure (31,35,37). One result of this was the observation of helically wound filamentous structures reminiscent of PHFs (31). The morphology of neurofilaments as seen in routine electron-microscopic sections is similar to the morphology of individual filaments forming the PHFs. However, as will be seen below, when both are viewed in negative stain, the infrastructure of the filaments making up PHFs and neurofilaments is quite different, which suggests that they are made of different proteins or assembled under conditions different from known cytoskeletal fibrils.

Isolated tangles of PHFs stained with either phosphotungstic acid (PTA) or uranyl acetate (UA) are readily seen at low magnifications (Figures 7-6 and 7-7). However, at higher magnifications (Figures 7-7 and 7-8), the two stains give quite different pictures of PHFs. Much greater detail is seen in the PTA preparations than those stained with UA. Both stains do show PHFs to be straight, unbranched profiles that give a rigid appearance. The negative-stain image of the PHFs is not the same as that observed in thin section. For example, in thin section, one can observe a separation between the filaments midway between the crossovers. In negative stain, on the other hand, there is no obvious separation. At high magnification, an infrastructure is apparent in the PTA-stained PHFs that is not seen in UA-stained preparations. As Figures 7-8, 7-9, and 7-10 show, each of the two filaments in a PHF is composed of two to four protofilaments, each of which is 3 to 6 nm in diameter. When they separate, the protofilaments appear to be helically wound around each other with a distance between crossover points of 40 nm, or half of that seen in a complete PHF. (See Figure 7-11). Tilt analysis, three-dimensional reconstruction, and metal-shadowing studies are now in progress to define these structures more clearly. Preliminary studies with tilt analysis revealed that the helical property of complete PHFs is retained in negatively stained preparations (Figure 7-12).

A comparative ultrastructural analysis has been done with negatively stained PHFs, neurofilaments, and microtubules. Neurofilaments can be isolated by axonal flotation or glycerol assembly from a variety of mammalian brains (19,21). We have examined neurofilaments stained with both UA and PTA (Figures 7-13 and 7-14). Unlike PHFs, glycerol-assembled neurofilaments are unstable in PTA and require prior fixation with glutaraldehyde. In contrast to the

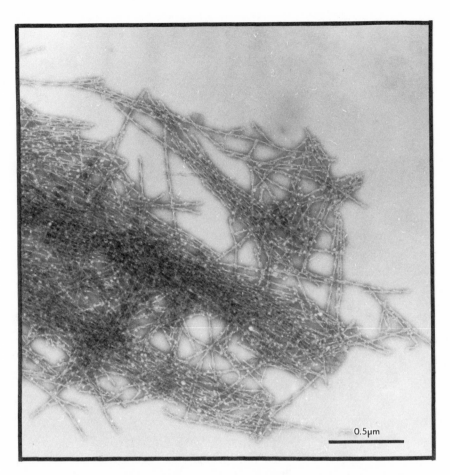

Figure 7-6. An aggregate of PHFs stained with PTA, demonstrating the typical appearance of PHFs in PTA-stained preparations. (Magnified ×60,000.)

straight protofilaments seen in PHFs and microtubules, the neurofilament protofilaments are difficult to resolve and are seen only in certain segments. Under optimal conditions, one can see that a neurofilament is composed of at least four protofilaments in a braided, ropelike arrangement (Figures 7-13 and 7-14).

Polymerized microtubules are also more readily stained with UA and require prior fixation with glutaraldehyde prior to staining with PTA (Figures 7-15 and 7-16). These microtubules share the rigid appearance of the PHFs, but their images, dimensions, and staining properties are different. At high magnification, the protofilaments are easily seen in tubulin sheets, ribbons of microtubules, and single microtubules (Figure 7-16). Again, as stated by others (1), these

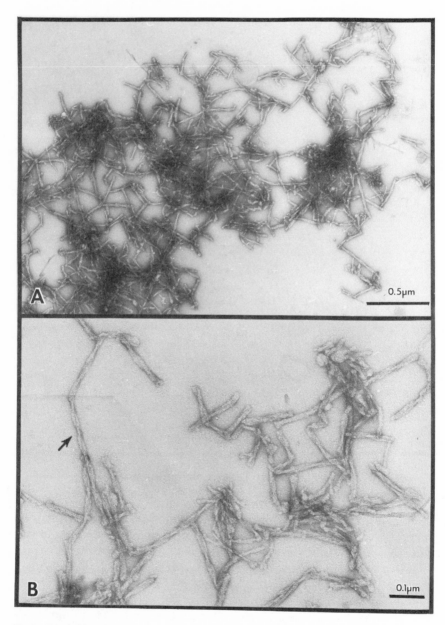

Figure 7-7. (A) Typical appearance of sonicated PHFs stained with UA. (Magnified ×48,000.) (B) PHFs, at higher magnification, illustrating the helical nature of the PHFs (arrow) and the lack of apparent infrastructure. Compare this with the UA-stained neurofilaments and microtubules (Figures 7-14B and 7-16B). (Magnified ×133,000.)

protofilaments are 4 to 5 nm in diameter. They, too, differ from PHF protofilaments in their staining properties and spacing. Thus, in

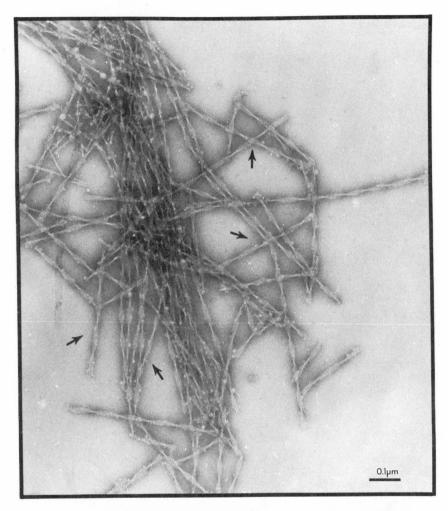

Figure 7-8. PHFs stained with PTA, illustrating the infrastructure apparent at medium magnification (arrows). Compare this figure with PTA-stained neurofilaments (Figures 7-13B) and microtubules (Figure 7-15B). (Magnified × 133,000.)

negative-stained preparations, microtubules (*in vitro* polymerized) are quite dfferent from PHFs.

The substructure of PHFs is more closely related to amyloid fibrils, particularly Alzheimer amyloid fibrils, than to any of the normal neurofibrils (32). The congophilic, green, birefringent fibrils such as the amyloids have a rigid appearance and, like PHFs, are more readily stained with PTA (14). X-ray diffraction and circular dichroism studies have shown that many of the amyloid fibrils are β-pleated. The observed negative-stain structure of the PHFs may reflect a similar β-pleated configuration. This awaits

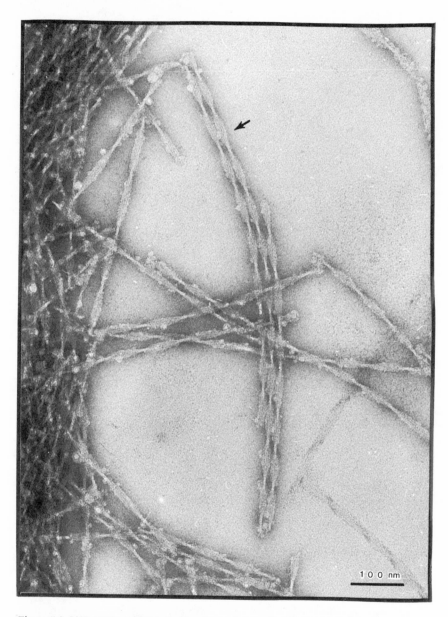

Figure 7-9. PHF tangle at high magnification and stained with PTA, revealing the apparent four protofilaments in almost every PHF (arrow). The protofilaments measure 3 to 6 nm. The PHFs measure 16 to 20 nm in diameter, narrowing every 80 nm to 6 to 10 nm. Compare this with normal neurofilaments (Figure 7-13B) and microtubules (Figure 7-15B). (Magnified ×221,000.)

confirmation with X-ray diffraction and circular dichroism studies. A variety of immunological studies indicate a possible relationship

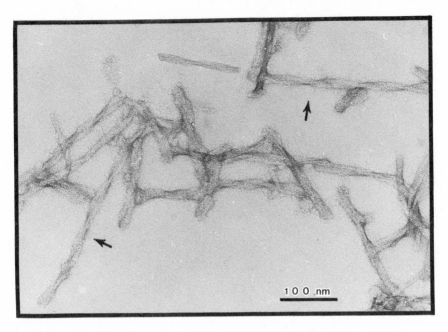

Figure 7-10. Sonicated PHFs stained with UA at high magnification, illustrating the lack of apparent infrastructure, the two-filament nature of the PHFs, and their helical nature (arrows). Compare this with normal neurofilaments (Figure 7-14B) and microtubules (Figure 7-16B). (Magnified ×211,000.)

between a component of cytoskeletal proteins and the PHFs (see below). However, it is clear that such a relationship is not apparent from their comparative ultrastructure in negatively stained preparations. A very promising approach to this issue is the ultrastructural analysis of antibody localization on PHFs.

Biochemical Characterization of Alzheimer Tangles and Paired Helical Filaments

PHF enriched fractions have been prepared by Iqbal and others (23, 25) through first isolating neuronal perikarya from Alzheimer brain under procedures developed by Norton and Poduslo (33) and modified by Iqbal and Tellez-Nagel (20) and then subjecting the isolated perikarya to a subcellular fractionation scheme. Polyacrylamide gel electrophoresis (PAGE) of the PHF enriched fractions revealed a major polypeptide at about 50,000-mol wt (PHF-P). This band was not seen in identically analyzed normal human brain tissue and was much diminished in fractions obtained from minimally affected regions of Alzheimer brain.

To establish possible relationships between PHF-P and other normal filamentous structures, Iqbal and others (18) compared the

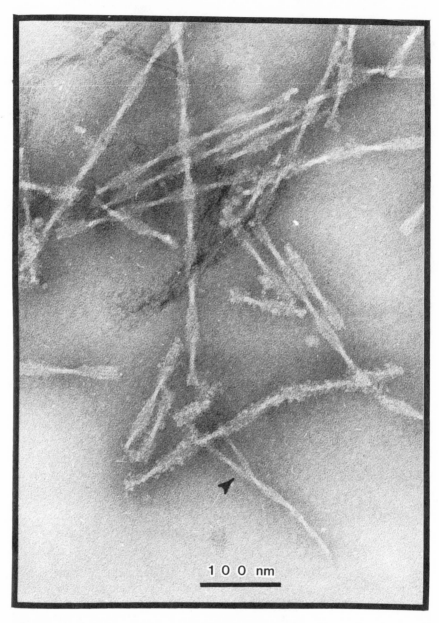

Figure 7-11. Sonicated PHFs stained with PTA showing (point) what appears to be a separation of a PHF into two filaments approximately 8 nm in diameter. Each 8-nm filament is composed of two (2- to 4-nm) protofilaments that also appear helically wound around each other with crossover points at about 40 nm (Magnified ×211,000.)

tryptic digestion maps of PHF-P, α-tubulin, β-tubulin, and what at the time was presumed to be a 50,000-mol wt neurofilament protein (P50). The PHF-P maps showed that (1) 22 of 24 P50 fragments were

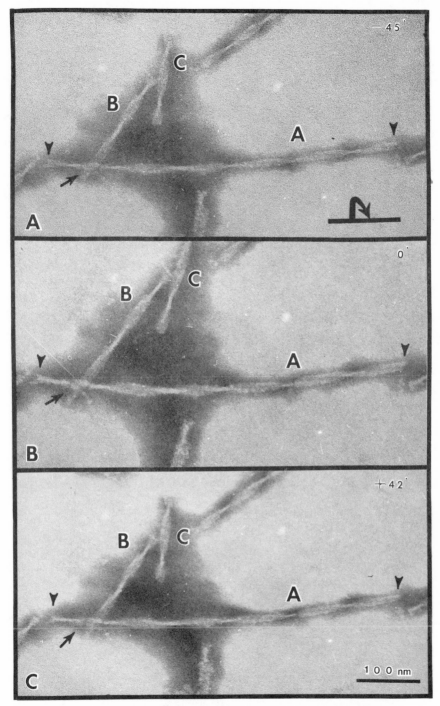

Figure 7-12. An illustration of PHFs tilted from −45° to +42° with a Hitachi 800. Filament A is parallel and is being rolled. Filaments B and C, approximately perpendicular to the axis of rotation, are being pitched. The apparent movement of the narrow portion of filament A to the left can be observed (arrows). Note the changes observed at the ends of filament A (points). (Magnified ×240,000.)

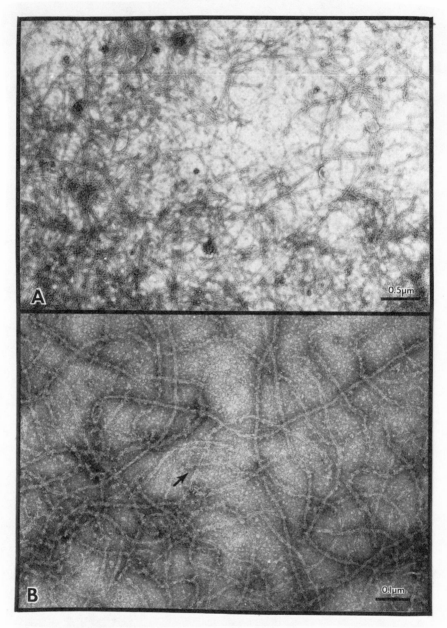

Figure 7-13. (A) An example of rabbit neurofilaments isolated by axonal flotation at low magnification (×30,000). (B) An example of rabbit neurofilaments isolated by glycerol at high magnification. (Magnified ×133,000.) In certain areas, the neurofilament substructure can be observed (arrow). A and B are stained with PTA after prior fixation with glutaraldehyde.

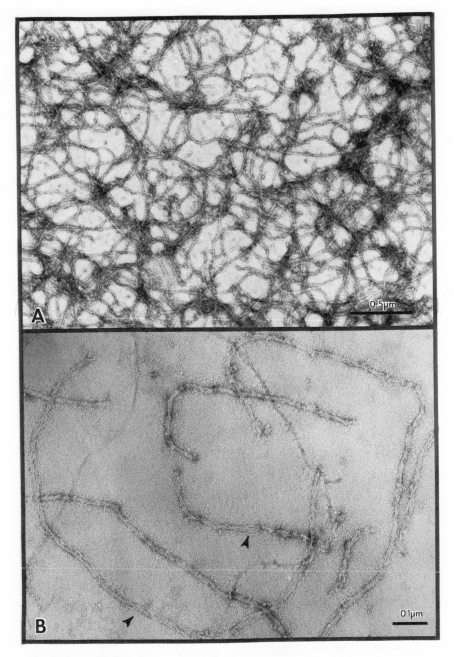

Figure 7-14. (A) An example of beef spinal cord neurofilaments isolated with glycerol and stained with UA. (Magnified ×48,000.) (B) Note that the neurofilament proto-filaments are rather difficult to resolve and are only seen clearly in certain segments (points). (Magnified ×133,000.)

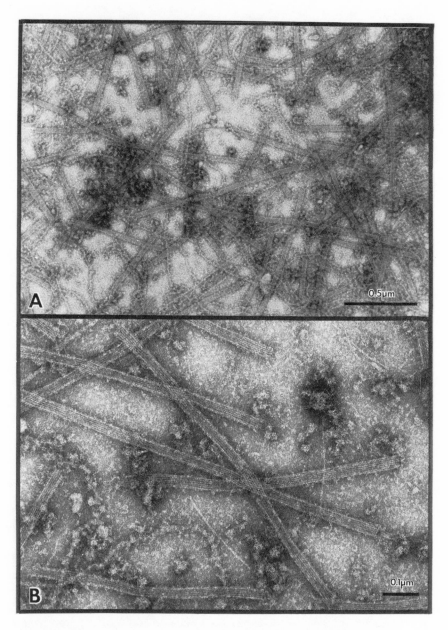

Figure 7-15. Examples of *in vitro* polymerized microtubules from beef brain, stained with PTA after prior fixation with glutaraldehyde. (A) Magnified ×57,000. (B) Higher magnification illustrates more clearly the microtubule protofilaments. (Magnified ×133,000.)

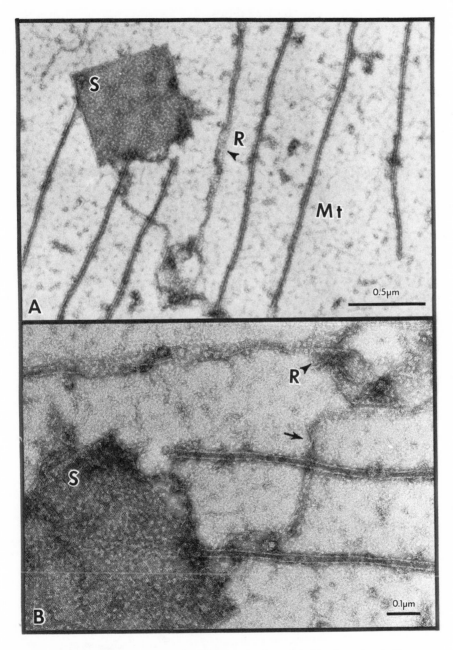

Figure 7-16. An example of the variety of structures formed during glycerol assembly of microtubles: tubulin sheets (S), ribbons of protofilaments (R), and microtubules (Mt) can be observed in *in vitro* polymerized microtubules prepared from human autopsy brain stained with UA. (A) Magnified ×60,000. (B) Illustration of the protofilaments in the tubulin sheet, ribbon, an observed twist (arrow), and the protofilaments observed as complete microtubules. (Magnified ×133,000.)

common to PHFs, (2) 18 of 23 β-tubulin fragments were common to PHF, and (3) 14 of 24 α-tubulin fragments were common to PHF (suggesting a close relationship between PHF-P, P50, and β-tubulin and a relatively weak relationship to α-tubulin) (50).

The identification of a more precise relationship between PHFs and other filamentous proteins in the neuron must await the purification and characterization of the constituent PHF polypeptides. So far, this has proven difficult, largely because of the physical properties of the PHFs themselves. The PHFs appear to be very stable structures. For example, their morphology or isolation characteristics do not apper to be affected by postmortem changes in the brain even over a wide range of times between death and autopsy. It is this stability that is at least partly responsible for the isolation of highly enriched PHF preparations (24,38). Unfortunately, this stability carries with it a very high degree of insolubility such that the solubilization of PHFs for gel electrophoresis is extremely difficult. In a recent report, Selkoe and others (38) concluded that even after heating PHFs to 100°C for 2 minutes in 2% sodium dodecyl sulfate (SDS), 5% β-mercaptoethanol (BME), 9.5-M urea, and 1% NP40, there were no polypeptide bands that (1) were unique to Alzheimer brain (as compared to identically treated normal human brain) or (2) were increased over control preparations. Electron-microscopic examination of the proteinaceous material that did not enter the gel revealed abundant PHFs. Recently, we (24) were able to show that repeated cycles of heating PHFs in 10% SDS and 10% BME to 100°C for 10 minutes resulted in (1) the disappearance of PHFs (as followed electron microscopically) in pellets after each cycle of heating and (2) the presence of several polypeptide bands in the 45,000- to 60,000-mol wt range as well as a high-molecular-weight band just below the top of the resolving gel. Electron microscopy of this high-molecular-weight fraction revealed no PHFs. It was also found that, when either the excluded protein, the high-molecular-weight protein, or the 45,000- to 60,000-mol wt polypeptides were eluted from the gel and electrophoresed again separately, each one generated the entire original pattern. Thus, it appears that considerable reaggregation may be occurring in the system.

The situation is also complicated by the fact that there is evidence suggesting that PHFs and/or tangles are associated with what appear to be nonconstituent proteins: for example, albumin and gamma globulin (47), prealbumin (40), and nucleoside phosphatase (27).

Immunohistochemistry of Alzheimer Tangles and Paired Helical Filaments

Of the antisera that have stained Alzheimer tangles, only one used PHFs as the antigen source (16). In this case, the antigen was the

PHF-P produced by Iqbal and others (25). The antiserum formed Ouchterlony precipitation lines not only with PHF-P but also with microtubules prepared from normal human brain by two cycles of assembly-disassembly. The antiserum also reacted with some neurofilament preparations but not with others. This same antiserum immunostained tangles in sections of Alzheimer brain both by fluorescence and PAP techniques. Congo red counterstaining showed that, although most tangles were stained, others clearly were not. The PAP staining also revealed a reaction at the margins of some neuritic plaques that was not seen in their cores. Presumably, this is a reflection of the presence of PHFs in the dystrophic neurites. Grundke-Iqbal and others also produced an antiserum against normal human microtubules purified by two cycles of assembly-disassembly (17). In an Ouchterlony double-diffusion system, the antiserum reacted with PHF-P, microtubules, and tubulin but did not react with neurofilament preparations. The antiserum also labeled tangles in brain sections and isolated perikarya from Alzheimer brain. Preadsorption of this antiserum with purified tubulin did not abolish the staining of the Alzheimer tangles (22). Yen and others have also reported an antiserum to human brain microtubules that stained Alzheimer tangles (54). In the same study, an array of other antisera to calf brain microtubules, sea urchin microtubules, beef brain microtubules, and the P200 neurofilament polypeptide failed to stain tangles. On the other hand, the antiserum to human brain microtubules also reacted with neurofilament triplet polypeptides, ferritin, and tubulin. However, studies with absorption and other specific antisera suggested that none of these proteins was responsible for the immunostaining of the tangles. Thus, both the Grundke-Iqbal and Yen studies revealed that there is an unidentified "micro-component" of microtubule fractions that is also present in Alzheimer neurofibrillary tangles.

Other immunostaining studies have shown that neurofibrillary tangles also contain what appear to be ill-defined neurofilament antigens. Ishii and others (26) produced an antiserum against a 50,000-mol wt "neurofilament" fraction that stained tangles. The serum also contained anti-glial filament acidic protein (GFAP) since it also stained astrocytes and their processes. This is consistent with the observation that neurofilament preparations of this type contain large amounts of GFAP and smaller amounts of what are probably degraded neurofilament proteins (5,7,53). Antiserum to a 50,000-mol wt neurofilament preparation from chicken brain has also been shown to stain tangles (6,13). Other studies with this serum show that it also reacts with each of the neurofilament triplet proteins and stains neurofilaments *in situ* (9). It also stains experimentally induced tangles (8). Recently, Anderton and others (2) reported that monoclonal antibodies to the 200,000-mol wt and the 70,000-mol wt neurofilament proteins also stain Alzheimer neurofibrillary tangles.

Finally, there are two recent reports showing that Alzheimer tangles are also immunostained by anti-prealbumin (40), anti-gamma globulin, and anti-albumin (47) and by a nonabsorbable component in an antiserum raised against vasoactive intestinal polypeptide (VIP) (28).

Conclusions

From the foregoing it is clear that the identity and origins of the Alzheimer neurofibrillary tangles and their PHFs are far from resolved. It is certainly tempting to regard them as being composed at least in part of degraded neurofilament protein. The protein's presence in microtubule preparations would also provide an explanation for the cross-reacting antigens reported by Grundke-Iqbal and associates (17) and Yen and associates (54). An intriguing aspect of this notion, however, is the fact that these postulated constituent proteins of PHFs are found in brains from humans (17), chickens (6), rats (26), and cattle (53); yet, the structure of interest, the Alzheimer-type PHF, is unique to human pathological conditions.

Although the connection to neurofilament degradation products is the current best guess, reports of the presence of serum proteins in tangles serve to raise an important caveat, namely, the need to be able to distinguish between the structural proteins of PHFs and the adventitious proteins that have become entrapped in the tangle and decorate the PHFs. One should also bear in mind that the neurofibrillary tangle may be composed of not only PHF but of 10-nm neurofilaments and 15-nm straight filaments similar to those described in supranuclear palsy as well (39,49). In addition, when studied *in situ*, PHFs appear to be in a proteinaceous matrix that might generate immunological reactions with unpolymerized cytoskeletal proteins. Thus, it appears that immunohistochemistry at the level of the light microscope is not likely to provide the definitive answer. The solution of the PHF problem must await the solubilization of the peptides, their isolation via immunoaffinity employing monoclonal antibodies, and the immuno-ultrastructural localization of these polypeptides on isolated PHFs.

REFERENCES

1. Amos, F. A. 1979. Structure of microtubules. *In*: Microtubules. K. Robert and J. S. Hyans, editors. Academic Press, London, pp. 1-64.

2. Anderton, B. H., D. Breinberg, M. J. Downes, P. J. Green, B. E. Tomlinson, J. Ulrich, J. N. Wood, and J. Kahn. 1982. Monoclonal antibodies show that neurofibrillary tangles and neurofilaments share antigenic determinants. *Nature* 298:84-86.

3. Bowen, D. M. 1980. Biochemical evidence for nerve cell changes in senile dementia. *In*: Aging of the Brain and Dementia. L. Amaducci, A. N. Davison, and P. Antuono, editors. Raven Press, New York, pp. 127-138.

4. Crapper-McLachlan, D. R., and U. De Boni. 1980. Etiologic factors in senile dementia of the Alzheimer type. *In*: Aging of the Brain and Dementia. L. Amaducci, A. N. Davison, and P. Antuono, editors. Raven Press, New York, pp. 173-181.

5. Dahl, D. 1981. Isolation of neurofilament proteins and of immunologically active neurofilament degradation products from extracts of brain, spinal cord and sciatic nerve. *Biochim. Biophys. Acta* 668:299-306.

6. Dahl, D., and A. Bignami. 1977. Preparation of antisera to neurofilament protein from chicken brain and human sciatic nerve. *J. Comp. Neurol.* 176:645-657.

7. Dahl, D., and A. Bignami. 1979. Astroglial and axonal proteins in isolated brain filaments: I. Isolation of the glial fibrillary acidic protein and of an immunologically active cyanogen bromide peptide from brain filament preparations of bovine white matter. *Biochim. Biophys. Acta* 578:305-316.

8. Dahl, D., A. Bignami, N. T. Bich, and N. H. Chi. 1980. Immunohistochemical characterization of neurofibrillary tangles induced by mitotic spindle inhibitors. *Acta Neuropathol. (Berl.)* 51:165-168.

9. Dahl, D., D. J. Selkoe, R. T. Pero, and A. Bignami. 1982. Immunostaining of neurofibrillary tangles in Alzheimer's senile dementia with a neurofilament antiserum. *J. Neurosci.* 2:113-119.

10. Davies, P., and A. J. F. Maloney. 1976. Selective loss of cholinergic neurons in Alzheimer's disease. *Lancet* 2:1403.

11. De Boni, U., and D. R. Crapper. 1978. Paired helical filaments of the Alzheimer type in cultured neurons. *Nature* 271:566-568.

12. Foelix, R. F., and M. Hauser. 1979. Helically twisted filaments in giant neurons of the whip spider. *Eur. J. Cell Biol.* 19:303-306.

13. Gambetti, P., H. P. Velasco, D. Dahl, A. Bignami, U. Roessmann, and S. D. Sindley. 1980. Alzheimer neurofibrillary tangles: An immunohistochemical study. *In*: Aging of the Brain and Dementia. L. Amaducci, R. N. Davison, and P. Antuono, editors. Raven Press, New York, pp. 55-68.

14. Glenner, G. G., C. D. Eanes, H. A. Bladin, R. P. Finke, and J. Termino. 1974. β-pleated sheet fibrils: A comparison of native amyloid with synthetic protein fibrils. *J. Histochem. Cytochem.* 22:1141-1158.

15. Grundke-Iqbal, I., K. Iqbal, P. Merz, and H. M. Wisniewski. 1981. Isolation and properties of Alzheimer neurofibrillary tangles. *J. Neuropathol. Exp. Neurol.* 40:312 (abs.).

16. Grundke-Iqbal, I., A. B. Johnson, R. D. Terry, H. M. Wisniewski, and K. Iqbal. 1979. Alzheimer neurofibrillary tangles: Antiserum and immunohistological staining. *Ann. Neurol.* 6:532-537.

17. Grundke-Iqbal, I., A. B. Johnson, H. M. Wisniewski, R. D. Terry, and K. Iqbal. 1979. Evidence that Alzheimer neurofibrillary tangles originate from neurotubules. *Lancet* 1:579-580.

18. Iqbal, K., I. Grundke-Iqbal, H. M. Wisniewski, and R. D. Terry. 1978. Chemical relationship of the paired helical filaments of the Alzheimer's dementia to human normal neurofilaments and neurotubules. *Bran Res.* 142:321-332.

19. Iqbal, K., P. Merz, and H. M. Wisniewski. 1981. Isolation of mammalian CNS neurofilaments by *in vitro* assembly-disassembly. *Trans. Am. Soc. Neurochem. 12*:200 (abs.).

20. Iqbal, K., and I. Tellez-Nagel. 1972. Isolation of neurons and glial cells from normal and pathological human brains. *Brain Res. 45*:296-301.

21. Iqbal, K., and H. Wisniewski. 1979. Neurofilament proteins: Effect of ionic strength and calcium chelation. *Trans. Am. Soc. Neurochem. 10*:168 (abs.).

22. Iqbal, K., I. Grundke-Iqbal, A. B. Johnson, and H. M. Wisniewski. 1980. Neurofibrillary proteins in aging and dementia. *In:* Aging of the Brain and Dementia. R. N. Davison and P. Antuono, editors. Raven Press, New York, pp. 39-48.

23. Iqbal, K., H. M. Wisniewski, I. Grundke-Iqbal, J. K. Korthals, and R. D. Terry. 1975. Neurofibrillary tangles of Alzheimer's presenile-senile dementia. *J. Histochem. Cytochem. 23*:563-569.

24. Iqbal, K., H. M. Wisniewski, and P. A. Merz. 1982. Alzheimer neurofibrillary tangles: solubilization of isolated paired helical filaments (in preparation).

25. Iqbal, K., H. M. Wisniewski, M. L. Shelanski, S. Brostoff, H. B. Liwnicz, and R. D. Terry. 1974. Protein changes in senile dementia. *Brain Res. 77*:337-343.

26. Ishii, T., S. Haga, and S. Tokutake. 1979. Presence of a neurofilament protein in Alzheimer's neurofibrillary tangles (ANT). *Acta Neuropathol. 48*:105-112.

27. Johnson, A. B., and N. R. Bloom. 1970. Nucleoside phosphatase activities associated with the tangles and plaques of Alzheimer's disease: A histochemical study of natural and experimental neurofibrillary tangles. *J. Neuropathol. Exp. Neurol. 29*:463-478.

28. Johnson, A. B., S. A. Cohen, S. I. Said, and R. D. Terry. 1981. Neuritic plaque amyloid, microangiopathy and Alzheimer neurofibrillary tangles: Do they share a common antigen? *J. Neuropathol. Exp. Neurobiol. 40*:310 (abs.).

29. Katzman, R. 1976. The prevalence and malignancy of Alzheimer's disease. *Arch. Neurol. 33*:217-218.

30. Kidd, M. 1963. Paired helical filaments in electron microscopy in Alzheimer disease. *Nature 197*:192-193.

31. Krishnan, N., I. R. Kaiserman-Abramof, and R. J. Lasek. 1979. Helical substructure of neurofilaments isolated from *Myxicola* and squid giant axons. *J. Cell Biol. 82*:323-335.

32. Merz, P. A., R. A. Somerville, and H. M. Wisniewski. 1981. Abnormal fibrils in scrapie and senile dementia of the Alzheimer type. Paper presented at the Symposium on Non-Conventional Viruses and Effects on the Central Nervous System, Paris, November 5-7.

33. Norton, W. T., and S. E. Poduslo. 1970. Neuronal soma and whole neuroglia of rat brain: A new isolation technique. *Science 167*:1144-1145.

34. Perry, E. K., R. H. Perry, G. Blessed, and B. E. Tomlinson. 1977. Necropsy evidence of central cholinergic deficits in senile dementia. *Lancet 1*:189.

35. Schlaepfer, W. 1977. Immunological and ultrastructural studies of neurofilaments isolated from rat peripheral nerve. *J. Cell Biol. 74*:226-240.

36. Schlaepfer, W. 1977. Studies on the isolation and substructure of mammalian neurofilaments. *J. Ultrastruct. Res. 61*:149-157.

37. Schlaepfer, W. 1979. Nature of mammalian neurofilaments and their breakdown by calcium. *In:* Progress in Neuropathology. Vol. 4. H. M. Zimmerman, editor. Raven Press, New York, pp. 101-123.

38. Selkoe, D. J., Y. Ihara, and F. J. Salazar. 1982. Alzheimer's disease. Insolubility of partially purified paried helical filaments in sodium dodecyl sulfate and urea. *Science 215*:1243-1245.

39. Shibayama, H., and J. Kitoh. 1978. Electron microscopic structure of Alzheimer's neurofibrillary changes in a case of atypical senile dementia. *Acta Neuropathol. 41*:229-234.

40. Shirahama, T., M. Skinner, P. Westermark, A. Rubinow, A. S. Cohen, A. Brun, and T. L. Kemper. 1982. Senile cerebral amyloid: Prealbumin as a common constituent in the

neuritic plaque, in the neurofibrillary tangle, and in the microangiopathic lesion. *Am. J. Path. 107*:41-50.

41. Struble, R. G., L. C. Cork, P. J. Whitehouse, and D. L. Price. 1982. Cholinergic innervation in neuritic plaques. *Science 216*:413-415.

42. Tomlinson, B. E., G. Blessed, and M. Roth. 1970. Observations on the brains of demented old people. *J. Neurol. Sci. 11*:205-242.

43. Volk, B. 1980. Paired helical filaments in rat spinal ganglia following chronic alcohol administration: An electron microscopic investigation. *Neuropath. Appl. Neurobiol. 6*:143-153.

44. Whitehouse, P. J., D. L. Price, A. W. Clark, J. T. Coyle, and M. R. Delong. 1981. Alzheimer's disease: Evidence for a selective loss of cholinergic neurons in the nucleus basalis. *Ann. Neurol. 10*:122-126.

45. Whitehouse, P. J., D. L. Price, R. G. Struble, A. W. Clark, J. T. Coyle, and M. R. Delong. 1982. Alzheimer's disease and senile dementia: Loss of neurons in the basal forebrain: *Science 215*:1237-1239.

46. Wisniewski, H. M., B. Ghetti, and R. D. Terry. 1973. Neuritic (senile) plaques and filamentous changes in aged rhesus monkeys. *J. Neuropathol. Exp. Neurol. 32*:566-584.

47. Wisniewski, H. M., and P. B. Kozlowski. 1982. Evidence for blood-brain barrier changes in senile dementia of the Alzheimer type (SDAT). *Ann. N.Y. Acad. Sci.*, in press.

48. Wisniewski, H. M., H. K. Narang, and R. D. Terry. 1976. Neurofibrillary tangles of paired helical filaments. *J. Neurol. Sci. 27*:173-181.

49. Wisniewski, H. M., R. S. Sinatra, K. Iqbal, and I. Grundke-Iqbal. 1981. Neurofibrillary and synaptic pathology in the aged brain. *In*: Aging and Cell Structure. Vol. 1. J. E. Johnson, editor. Plenum Press, New York, pp. 104-142.

50. Wisniewski, H. M., and D. Soifer. 1979. Neurofibrillary pathology: Current status and research prospectives. *Mech. Aging Devel. 9*:119-142.

51. Wisniewski, H. M., and R. D. Terry. 1976. Neuropathology of the aging brain. *In*: Neurobiology of Aging. R. D. Terry and S. Gershon, editors. Raven Press, New York, pp. 265-280.

52. Wisniewski, K., G. A. Jervis, R. C. Moretz, and H. M. Wisniewski. 1979. Alzheimer neurofibrillary tangles in diseases other than senile and presenile dementia. *Ann. Neurol. 5*:288-294.

53. Yen, S.-H., D. Dahl, M. Schachner, M., and M. L. Shelanski. 1976. Biochemistry of filaments of the brain. *Proc. Natl. Acad. Sci. USA. 73*:529-533.

54. Yen, S.-H., F. Gaskin, and R. D. Terry. 1981. Immunocytochemical studies of neurofibrillary tangles. *Am. J. Path. 104*:77-89.

Chapter 8

Some Aspects
of Current and Future
Neurofilament Research

Charles A. Marotta

Significant aspects of neuronal filament research have been described in the previous chapters of this volume. This brief section will serve only to highlight those areas that may prove to be particularly fruitful for increasing our knowledge of the structure, function, and genetic regulation of neurofilaments.

Considerable effort has been invested in identifying the major protein constituents of neurofilaments (NFs). Modified versions of the axon-flotation procedure of Norton and associates (Chapter 2) or the direct extraction procedure of Schlaepfer (Chapter 3) have been used to prepare neurofilaments for protein analysis. Various chapters of this volume have described the evidence that proteins of approximately 200,000-, 145,000-, and 68,000-mol wt are the major constituents of neurofilaments. However, the initial biochemical and immunological identification was guided by the axoplasmic transport studies of Hoffman and Lasek (10). In their pioneering investigations, these authors observed the synchronous movement of three proteins in the slowest component of axoplasmic transport, and they hypothesized that these proteins comprise neurofilaments.

Recent evidence that the triplet proteins are neurofilament proteins (NFPs) can be briefly summarized as follows: (i) proteins of approximately the same size as those described by Hoffman and Lasek are among the major proteins extracted from CNS and PNS axons (Chapters 2 and 3); (ii) antibodies to each of the three proteins bind to neurofilaments *in situ* and *in vitro* (23,28); (iii) as neurofilaments

This chapter was completed while the author was the recipient of NIH research career development award AG00084 and a MacArthur Foundation award.

disintegrate in lesioned nerves or in nerves exposed to exogenous calcium, the triplet proteins are lost concomitantly (Chapter 3); (iv) as neurofilaments accumulate in nerves treated with aluminum or β,β'-iminodiproprionitrile, the triplet proteins increase concurrently (Chapter 6). These data do not rule out the possibility, however, that other proteins, in addition to the triplet (e.g., enzymes of phosphate metabolism [13,22,23,25], are also intimately associated with neurofilaments and contribute to their morphology or functional capacity.

More recent studies have begun to appreciate the unique contributions of the individual NFPs to the filamentous fine structure. Using antibodies to each of the purified proteins, Willard and Simon (28) established the fact that, although the three proteins may be physically associated with an individual neurofilament, their respective locations were nonidentical. The smallest subunit (73,000-mol wt) appeared to form a helical core, whereas the largest subunit (195,000-mol wt) was peripherally attached and periodically arranged along the axis. In some cases the latter protein appeared to form cross-bridges between filaments and in other cases appeared as a helix wrapping the central core. (See Chapter 4.)

The suggestion of Willard and Simon that the smallest NFP is a core protein has been supported by neurofilament reconstitution experiments. Moon and others (16) showed that under specified conditions bovine brain filaments were assembled from a complex of 160,000- and 78,000-mol wt NFPs previously suspended in 8-M urea. The filaments obtained were up to 500 nm in length; however, the formation of longer filaments required the presence of the 200,000-mol wt NFP. Zackroff and others (29) reassembled bovine spinal cord neurofilaments from the triplet proteins previously solubilized in low salt or 8 M urea. However, only the 68,000-mol wt protein was capable of reassembly into 10-nm filaments in the absence of other NFPs. In a separate study (9), it was also shown that porcine spinal cord 68,000-mol wt NFP, previously solubilized in 6-M guanidine hydrochloride, self-assembled into smooth 8- to 10-nm filaments. Intermediate filaments could not be reconstituted from the 160,000-mol wt or 200,000-mol wt NFPs alone; however, addition of these species to the 68,000-mol wt protein during self-assembly led to the formation of shorter filaments with rough surfaces containing whiskerlike protrusions along the axis. More recently, Liem and Hutchison (13) definitively demonstrated that the 70,000-mol wt protein by itself is capable of forming an intermediate filament whereas the other two proteins are not.

Although it appears that the subunit proteins contain the primary structural information for formation of intermediate filaments (9), *in vivo* this process may be influenced by local ionic conditions, the availability of energy sources, and the presence of cofactors. Thus, more extensive *in vitro* reconstitution experiments would be useful in defining the detailed molecular mechanism of filament formation.

Electron microscopy data have shown that the filaments assembled from the 68,000-mol wt NFP to have a periodicity of 21 nm, similar to self-assembled desmin and keratin (9,11). This length can be accommodated by the model for intermediate filaments proposed by Steinert and associates (25,26). The common structural unit was described as three intertwined polypeptide subunits with two discrete coild-coil α-helical segments, each 18 nm long, interspersed with nonhelical domains of 4 nm. Elongated filaments would consist of many units associated end to end and side by side.

Recent amino acid sequence analysis studies lend further support to the concept that the 68,000-mol wt NFP is the neuronal equivalent of intermediate filament protein found in other cell types. Weber and associates (8) determined the sequence of a 44-residue fragment from the 68,000-mol wt protein. Homologies were established between this neurofilament protein and similar fragments from vimentin and desmin (11). Greater than 40% of the amino acids were shown to be identical among the three proteins.

Thus, data derived from antibody-decoration experiments, *in vitro* reassembly studies, and amino acid sequence analyses support the view that the core neuronal intermediate filament protein is the smallest member of the triplet. Future studies will be necessary to elucidate the detailed contribution of each of the NFPs to the integrity of the neuronal cytoskeleton. In particular, it would be useful to establish whether or not the two larger NFPs are associated with neurofilament sidearms and whether or not these proteins have a specific role in neurofilament-microtubule interactions *in vivo* (See Chapter 2.)

A number of immunological investigations have been invaluable in demonstrating the existence of shared and unique antigenic sites among the triplet proteins, the distribution of antigens in different neural tissues, and the relationship of neurofilaments to the neurofibrillary tangles of Alzheimer's disease. Aspects of these various topics are discussed in Chapters 2, 3, 4, and 7, although the scientific literature contains numerous contradictory claims. (For a review, see reference 15.) Inconsistent neuroimmunological results are probably related to the variable specificities of antineurofilament protein sera produced by different laboratories. Variations may be related to the following: (i) the contamination of NFPs used as antigens by other proteins, (ii) the physical state of the antigen (denatured versus nondenatured), (iii) the use of nonuniform immunization protocols, (iv) the use of different animal species for antibody production, (v) the presence of autoantibodies to intermediate filament proteins in preimmune serum, (vi) the sensitivity of the various immunoassay procedures, and (vii) the use of nonuniform procedures to prepare and process tissues for immunostaining. The last consideration is particularly important for studies utilizing postmortem human brain tissue (15). We (2,15) and others (23) have discussed several of these variables.

A number of methodological difficulties related to the use of polyclonal antibodies can be overcome with monoclonal antibodies. Recently, a monoclonal antibody was used to demonstrate the existence of a common antigen in all classes of intermediate filament proteins from vertebrate and invertebrate sources (20). Immunological studies with a monoclonal antibody to the 200,000-mol wt NFP established the existence of a subpopulation of neurons that lack a specific NFP antigen (7). Studies using monoclonal antibodies to the two larger NFPs applied to affected tissue from Alzheimer brain demonstrated that neurofibrillary tangles share at least two antigenic sites with neurofilaments (1). Although applicable to a host of investigations, monoclonal antibodies would be particularly useful for the detailed study of the embryological expression of various NFP antigens and their distribution in structurally or functionally distinct areas of the nervous system.

In Chapter 4, Willard presents a series of observations aimed at clarifying the relationship of neurofilaments to axonal transport. Among the topics considered is the molecular form in which filaments are transported. Are they transported as a polymerized neurofilament-microtubule lattice or as precursors to the lattice? Aspects of both models are considered in some detail. The moving precursor hypothesis appears particularly intriguing since this mode of transport would provide a supply of precursor materials for repair of the cytoskeleton in the event of injury. In Chapter 5, Nixon advances the concept that NFPs are subject to proteolysis within axons, as well as at synaptic terminals. The viability of these hypotheses would benefit from the demonstration that processing of neurofilament subunits occurs, if only to a slight degree, during axoplasmic transport. We recently presented evidence that the middle-sized NFP is posttranslationally processed during transport (19). In the mouse, this protein occurs in microheterogeneous forms of 145,000-, 143,000-, and 140,000-mol wt (2,3,18,19). Using radioactive precursors injected into retinal ganglion cells, an increase in 143,000- and 140,000-mol wt forms was observed concomitantly with a decrease in the 145,000-mol wt protein (19). Similar observations were made when phosphorylated forms of these proteins were studied (18). On the basis of these data, we can partly explain the microheterogeneity of the middle-sized NFP (3) as due to both posttranslational phosphorylation and limited proteolysis. The contribution of heterogeneous NFPs to the overall structure of neurofilaments is unknown. It is possible that subtle structural variations in individual triplet proteins can result in polymorphism among neurofilaments. Detailed investigations of neurofilament heterogeneity may eventually help to define a structural role for these organelles.

Molecular genetic studies on neurofilament proteins have lagged far behind morphological, chemical, and physiochemical investigations. No direct information is currently available on the number of genes involved in specifying the neurofilaments. However, peptide mapping and amino acid analyses have shed some early light on the relationship of the triplet proteins to one another and have led to tentative deductions concerning the organization of NFPs at the level of the gene.

Purified 200,000-, 160,000-, and 70,000-mol wt NFPs from bovine, rat, and human CNS were compared by limited digestion with *Staphylococcus aureus* V8 protease or by cyanogen bromide cleavage (4). NFPs within the same size class but from different species gave similar peptides with both mapping techniques; however, the three size classes of NFPs of the same species gave different maps. The same results were obtained using rat and mouse NFPs (3). These studies led to the following conclusions: (i) mammalian NFPs in any one size class have similar structures, (ii) NFPs in different size classes are not related to one another by simple oligomerization, and (iii) the lower-molecular-weight NFPs are not derived from the higher. In Chapter 2, Chiu, Goldman, and Norton report the amino acid composition of the triplet proteins and show that each NFP has a unique amino acid composition. As a whole, the data support the involvement of separate genetic loci. The aforementioned report of Weber and associates (8) concerning a portion of the amino acid sequence of the 68,000-mol wt protein established homologies between this NFP and similar regions of vimentin and desmin. The results support the hypothesis that neuronal and nonneuronal intermediate filament proteins of the same size class are related in sequence and belong to a multigene family. Remaining for future exploration is the mechanism by which different family members are expressed and become associated with histologically distinct cell types.

Study of the *in vitro* synthesis of neurofilament proteins eventually may lead to more-refined investigations at the level of gene expression. At present, however, this experimentation faces technical obstacles. Spinal cord polysomes, translated in a homologous cell-free amino acid-incorporating system, were initially thought to be incapable of synthesizing NFPs; this idea led to speculation about the existence of precursor protein(s) (5). However, later studies (6) established that spinal cord free polysomes in a homologous protein-synthesizing system are capable of synthesizing polypeptides of the same size and isoelectric point as the triplet proteins. Unfortunately, other identifying criteria were not reported. An attempt made to synthesize the NFPs using spinal cord polysomes translated in a reticulocyte system led to the synthesis of a 68,000-mol wt polypeptide and trace amounts of 200,000- and 150,000-mol wt species (6). Spinal cord mRNA programmed into a wheat germ homogenate synthesized neither the 200,000-mol wt nor the 150,000-mol wt proteins, and an apparent degradation product (at 139,000 mol wt) was obtained (27). This may be attributed partly to the known high levels of endogenous proteases present in wheat germ (17) and to the well-established extreme sensitivity of neurofilament proteins to proteolytic degradation, as described in Chapters 3 and 5 of this volume. Although recent data have been interpreted as proving the existence of separate mRNAs (and hence separate genes) for the triad of NFPs, direct evidence is not yet available from protein synthesis experiments.

There is little doubt, however, that in the near future neurofilament proteins will be subjected to an increasing number of more sophisticated molecular biological investigations. The construction of DNA clones containing NFP sequences will allow us to approach answers to a number of significant questions concerning the genetic organization and regulation of neurofilaments. These questions concern the structural relationships among the genes, their numbers, and their chromosomal location; the evolutionary constraints on the different intermediate filament protein genes; the mechanisms of gene expression during embryogenesis, development and aging; and the possible occurence of genetic defects in diseases associated with abnormal filaments, such as Alzheimer's disease.

REFERENCES

1. Anderton, B. H., D. Breinburg, M. J. Downes, P. J. Green, B. E. Tomlinson, J. Ulrich, J. N. Wood, and J. Kahn, 1982. Monoclonal antibodies show that neurofibrillary tangles and neurofilaments share antigenic determinants. *Nature 298*:84-86.

2. Brown, B. A., R. E. Majocha, D. M. Staton, and C. A. Marotta. 1983. Axonal polypeptides crossreactive with antibodies to neurofilament proteins. *J. Neurochem. 40*:229-308.

3. Brown, B. A., R. A. Nixon, P. Strocchi, and C. A. Marotta. 1981. Characterizations and comparison of neurofilament proteins from rat and mouse CNS. *J. Neurochem. 36*:143-153.

4. Chiu, F.-C., B. Korey, and W. T. Norton. 1980. Intermediate filaments from bovine, rat and human CNS: Mapping analysis of the major proteins. *J. Neurochem. 34*:1149-1159.

5. Czosnek, H. H., D. Soifer, and H. M. Wisniewski. 1979. Biosynthesis of filamentous proteins by rabbit spinal cord: *in vivo* and *in vitro* studies. *Inter. Soc. Neurochem. (Abstr.) 7*:284.

6. Czosnek, H. H., D. Soifer, and H. M. Wisniewski. 1980. Studies on the biosynthesis of neurofilament proteins. *J. Cell Biol.* 85:726-734.

7. Debus, E., G. Flugge, K. Weber, and M. Osborn. 1982. A monoclonal antibody specific for the 200K polypeptide of the neurofilament triplet. *EMBRO J.* 1:41-45.

8. Geisler, N., U. Plessmann, and K. Weber. 1982. Related amino acid sequences in neurofilaments and nonneuronal intermediate filaments. *Nature* 296:448-450.

9. Geisler, N., and K. Weber. 1981. Self-assembly *in vitro* of the 68,000 molecular weight component of the mammalian neurofilament triplet proteins into intermediate-sized filaments. *J. Mol. Biol.* 151:565-571.

10. Hoffman, P. N., and R. J. Lasek. 1975. The slow component of axonal transport: Identification of major structural polypeptides of the axon and their generality among mammalian neurons. *J. Cell Biol.* 66:351-366.

11. Lazarides, E. 1980. Intermediate filaments as mechanical integrators of cellular space. *Nature* 283:249-256.

12. Leterrier, J.-F., R. K. H. Liem, and M. L. Shelanski. 1981. Preferential phosphorylation of the 150,000 molecular weight component of neurofilaments by a cyclic AMP-dependent, microtubule-associated protein kinase. *J. Cell Biol.* 90:755-760.

13. Liem, R. K. H., and S. B. Hutchison. 1982. Purification of individaul components of the neurofilament triplet: Filament assembly from the 70,000 dalton subunit. *Biochemistry* 21:3221-3226.

14. Liem, R. K. H., S.-H. Yen, G. D. Salmon, and M. L. Shelanski. 1978. Intermediate filaments in nervous tissue. *J. Cell Biol.* 79:637-645.

15. Marotta, C. A. 1983. Neuronal intermediate filaments. *In*: Handbook of Neurochemistry. A. Lajtha, editor. Plenum, New York (in press).

16. Moon, H. M., T. Wisniewski, P. Merz, J. DeMartini, and H. M. Wisniewski. 1981. Partial purification of neurofilament subunits from bovine brains and studies on neurofilament assembly. *J. Cell Biol.* 89:560-567.

17. Mumford, R. A., C. B. Pickett, M. Zimmerman, and A. W. Strauss. 1981. Protease activities present in wheat germ and rabbit reticulocyte lysates. *Biochem. Biophys. Res. Commun.* 103:565-572.

18. Nixon, R. A., B. A. Brown, and C. A. Marotta. 1981. Modification of neurofilament proteins in the mouse retinal ganglion cell axon. *Inter. Soc. Neurochem. (Abstr.)* 8:82.

19. Nixon, R. A., B. A. Brown, and C. A. Marotta. 1982. Post-translational modification of a neurofilament protein during axoplasmic transport: Implications for regional specialization of CNS axons. *J. Cell Biol.* 94:150-158.

20. Pruss, R. M., R. Mirsky, M. C. Raff, R. Thorpe, A. J. Dowding, and B. H. Anderton. 1981. All classes of intermediate filaments share a common antigenic determinant defined by a monoclonal antibody. *Cell* 27:419-428.

21. Runge, M. S., M. R. El-Maghrabi, T. H. Claus, S. J. Pilkis, and R. C. Williams, Jr. 1981. A MAP-2-stimulated protein kinase activity associated with neurofilaments. *Biochemistry* 20:175-180.

22. Runge, M. S., P. B. Hewgley, D. Puett, and R. C. Williams, Jr. 1979. Cyclic nucleotide phosphodiesterase activity in 10-nm filaments and microtubule preparations from bovine brain. *Proc. Natl. Acad. Sci. USA* 76:2561-2565.

23. Schlapefer, W. W., V. Lee, and H.-L. Wu. 1981. Assessment of immunological properties of neurofilament triplet proteins. *Brain Res.* 226:259-272.

24. Shelanski, M. L., J.-F. Leterrier, and R. K. H. Liem. 1981. Evidence for interactions between neurofilaments and microtubules. *Neurosciences Res. Prog. Bull.* 19:32-43.

25. Steinert, P. M. 1978. Structure of the three-chain unit of the bovine epidermal keratin filament. *J. Mol. Biol.* 123:49-70.

26. Steinert, P. M., W. W. Idler, and R. D. Goldman. 1980. Intermediate filaments of baby hamster kidney (BHK-21) cells and bovine epidermal keratinocytes have similar ultrastructures and subunit domain structures. *Proc. Natl. Acad. Sci. USA* 77:4534-4538.

27. Strocchi, P., D. Dahl, and J. M. Gilbert. 1982. The biosynthesis of intermediate filament proteins in the rat CNS. *Trans. Am. Soc. Neurochem. 13*:242.

28. Willard, M., and C. Simon. 1981. Antibody decoration of neurofilaments. *J. Cell Biol. 89*:198-205.

29. Zackroff, R. V., W. W. Idler, P. M. Steinert, and R. D. Goldman. 1982. *In vitro* reconstitution of intermediate filaments from mammalian neurofilament triplet polypeptides. *Proc. Natl. Acad. Sci. USA 79*:754-757.

Index

Index